ideaFest Journal

Volume 7
2023-2024

The Interdisciplinary Journal
of Creative Works and Research
from
Cal Poly Humboldt ideaFest

The Press at Cal Poly Humboldt

The Press at Cal Poly Humboldt publishes high-quality scholarly, intellectual, and creative works by or in support of our campus community. The press supports the Cal Poly Humboldt mission to improve the human condition and our environment by promoting understanding of social, economic, and environmental issues.

The Press at Cal Poly Humboldt
Cal Poly Humboldt Library
1 Harpst St.
Arcata, CA 95524-8299
digitalcommons.humboldt.edu

ISBN: 978-1-962081-02-3
ISSN: 2689-1891

Managing Editor and Layout : Thalia Barry
Design: Sarah Godlin
Additional Editing, Layout, and Design:
Alannah Guevara
Kiran Dunning
Tanya Gonzalez

ideaFest
Journal

Contents

IdeaFest
Journal

Reflective Perspective

Lisa Heikka-Huber, Cal Poly Humboldt

Abstract

'Reflective Perspectives' is a personal tribute to glass. After working in glass for over 20 years creating functional sculpted objects, I decided it was time to take a break and complete my BFA in sculpture. Influenced heavily from my sculptor parents, I sought to learn their craft and focus my art practice to combine metals with hollow glass sculpture. Taking a mold from my face frozen in the expression I make while blowing glass. Each bubble represents the constant influx of ideas for glass that flowed and formed in my brain throughout the years. Each idea started out as a spark of imagination. Some came to life, finding their way out into their final forms from my breath, in glass while others await realization in the future. Glass has consumed my mind for so long the transition out of the daily interactions with the medium turned out to be far more challenging than I expected but yielded invaluable time for development of my art practice. Cast in bronze, polished, patinaed, and then affixed with the glass elements, this piece represents my admiration for glass as a medium and a catalyst for my imagination. The color choice for the glass reflects the hue which I find most calming when life gets crazy and often use to facilitate personal level keel, a sort of mental centering purple. Hidden to the naked eye are the layers of ultraviolet pink color which coats this purple and comes to life under the black light, insinuating the hidden elements of my own imagination, reflected.

Keywords: Glass art, Bronze art, Metal casting, Interdisciplinary art, Mixed media art, Metal art, Glass art, Female artist, Female sculptor, Fine arts

Acknowledgements

Sincerest thanks to the Art Department at Cal Poly Humboldt for their continued support during my BFA especially Ben Funke, Sondra Schwetmann, Emily Cobb, Heather Madar, Erin Austin and Tova Lund. Additionally, thanks to my husband Bernd, daughter Marina and mother Suzanne for their encouragement to follow my dreams and complete my degree here in Humboldt.

Reflective Perspective, 2022
A self-portrait of the artist

Hidden ultraviolet pink color
layered atop glass elements

Borosilicate glass, bronze, patina
10"x7.5"x12"

IdeaFest
Journal

Approaches to Assessing Nutrient Coupling in Open Ocean Datasets

James M. Moore & Claire P. Till, Cal Poly Humboldt

Abstract

Nutrient coupling describes a process where the biogeochemical cycles of two elements are linked by being incorporated similarly into biomass. This paper uses data from the GEOTRACES GP16 cruise (Eastern Pacific Zonal Transect) to investigate the relationship between certain macronutrients generally coupled to trace elements in terms of their oceanic distributions with the notable exception of in an oxygen minimum zone: cadmium-phosphate and zinc-silicate. There are many methods applied to oceanographic data to correlate analyte concentrations; while they are often presented independently in literature, here we attempt to use them in conjunction for a more thorough interpretation. By compiling 1) depth profiles, 2) differences between the concentrations after applying a scale factor, and 3) direct comparisons of the analytes, greater insight into the correlations between coupled nutrients is achievable. Because the goal of studying this phenomenon is to learn about the underlying causes of these biogeochemical cycles, being able to characterize the decoupling in a way that is easy to visualize is extremely important to the efficacy of this area of research. Variation between results is elaborated on as a caution towards carefully considered and consistent data manipulation. Calculating a difference between the coupled analytes after applying a scaling factor provided the most useful information about decoupling.current lives are symbolic.

Keywords: Nutrient Coupling, Coupled Datasets, Biogeochemical Cycling, GEOTRACES, GP16, Trace Elements.

Introduction

The study of Earth's systems chemistry is critical to gaining a deeper understanding of the dynamics underlying air, water, and land masses. Earth maintains a complex network of concentration gradients spanning the entire globe, and our level of understanding of those biogeochemical cycles limits the capabilities of global mathematical models.

The GEOTRACES program is an international collaboration of scientists working to expand understanding of marine biogeochemical cycles by analyzing ocean samples for trace chemical constituents. GEOTRACES uses scientific research vessels to sample the ocean at a wide range of depths, longitudes, latitudes, and times to quantify the ocean's trace constituents. Along with trace element concentrations, isotope ratios can be used to identify and track water masses as well as physical and chemical processes.[1] The immediate result of a research cruise is a better understanding of the water chemistry along that specific transect of the ocean, but over a longer period, these kinds of datasets help elucidate the mechanisms controlling biogeochemical cycles as well and allow for substantial improvements to mathematical models of the ocean as a whole.[2–4] Ultimately, large-scale efforts like this benefit mankind because our ability to predict and control our environment on a large scale becomes increasingly feasible with more data and better Earth systems modeling.

Oceanographic data has been used to observe that some macronutrients are coupled to trace elements in terms of their oceanic distribution.2,5 This work addresses a few relationships between a macronutrient and a trace metal that have nutrient-type distributions evident in their depth profiles: cadmium-phosphate, zinc-silicate, and copper-silicate.2 Nutrient coupling can occur when two dynamic analyte concentration cycles are incorporated similarly into biomass.6 Bruland7 describes three main factors that influence whether a trace element will have a distribution correlated with a nutrient element: similar internal cycles, a long oceanic residence time relative to the time scale of ocean mixing, and that deep water scavenging is insignificant compared to the time scale of ocean mixing.

A deviation between coupled analytes from their expected values has been correlated to oxygen minimum zones (OMZs). OMZs are regions in the water column that have the relative least oxygen concentration, and severe OMZs often correspond to oxygen level below 5% saturation or < 20 µmol/L.8 There are rapid changes in the rates of uptake and remineralization between coupled analytes at depths correlated to OMZs, occurring before their difference in rates settle and the two analytes couple again at greater depths.5,9 The vertical depth profile of the OMZs show variation throughout the world, making the relationship between OMZs and nutrient coupling harder to isolate.10 Often OMZs lie below the euphotic zone in the range of approximately 100-1000 m, but Peru has an OMZ that partially reaches in the euphotic zones. The euphotic zone is described as a region receiving at least 1% the photosynthetically active radiation as the surface. Oxygen levels for OMZs in the Northeastern Pacific do not show a recovery in oxygen levels until 1000-1500 m deep though this Peru transect has its OMZ reported as stretching as low depths of > 3000 m.11 Resolution on the spatial distribution of the OMZ is limited by the constant fluctuations that come with time. For instance, periodic El Niño events have shown increased oxygenation and decreased nutrient availability due to downwelling, with more extreme fluctuations during equatorial Pacific El Niño events compared to central Pacific El Niño events, while the OMZ is also subject to non-periodic long-term changes such as the decrease in oxygen concentrations over the previous five decades.12

Reports for the GP16 transect show decoupling of their concentration profiles in the phototrophic zone by a loss of cadmium relative to phosphate2 and a loss of zinc relative to silicate13. There are several proposed explanations for this decoupling. One idea is that the decoupling is due to sulfide precipitation in low oxygen zones14; if cadmium precipitates out as CdS in a sulfidic environment, then it stands to reason that the decoupling would be observed in this sulfur rich and oxygen starved region. Middag et al. postulates that it is not the oxygen levels that cause the decoupling but a difference in biological remineralization rates which happens to be indirectly related to oxygen levels (due to the aerobic biota consuming oxygen).9 They supported this with data that showed the decoupling phenomenon occurring in higher oxygen waters (>75 µM), and they demonstrated a kink in the relationship between cadmium and phosphate could be observed by modeling the mixing of water masses only.9 Determining the exact cause of the decoupling is outside the scope of this paper; instead, herein, we focus on tools that can be used to assess coupling.

One difficulty faced when addressing this topic is the lack of standardization in the method of comparing coupled elements. Currently, common methods to compare relationships between two elements are to plot the difference between the values scaled appropriately,2,15,16 to plot the concentration of one element against another,7–9,15,17–23 or to use a spot test that tabulates average ratios of the analytes for different water masses.9,15 Each comparison has its own benefits and drawbacks, and all are valuable; the key aspect is to select the comparison method intentionally and with full understanding of its significance. Ideally, the same comparison method would be the go-to approach across different regions and transects, to allow for broader conclusions to be more easily made.

This paper is a review of approaches, with commentary, and a suggestion of a new approach using the data from the GEOTRACES Eastern Pacific Zonal Transect (EPZT) GP16 cruise to provide real world insight on how the different comparison methods extract information from the cruise sample data. Stations 1-17 were used in this study which stretch from Peru to Tahiti, 80 W to 150 W roughly along the latitude 12 S. The transect passes right through one of the world's most extreme OMZs, the OMZ off Peru, reported to be between 80 to 90 W and 0 to 18 S.11 These data are used to justify the methods researchers have used to report the extent of decoupling and distinguishes between the different methods.

Methods

Sample and Instrumentation Procedure

Samples from GP16 were collected following GEOTRACES protocols24 from the carousel of Teflon-coated 12 L Go-Flo bottles, filtered through 0.2-micron supor acropak capsule filters, and acidified at sea with 4 mL of 6 M quartz-distilled HCl per liter of seawater. Samples were analyzed in the Bruland lab at UC Santa Cruz after sitting for several months acidified. The analytical method is described in Biller and

Bruland25 with adaptations as in Parker et al.26 Briefly, this method involves offline pre-concentration on Nobias chelate PA1 resin at pH 6.0 +/- 0.2, followed by analysis on the high resolution, extended range, Thermo Element inductively coupled plasma mass spectrometer (HR ICP-MS). All concentration data used here underwent intercalibration and are available at the GEOTRACES Intermediate Data Product.3

Plots Used for Investigating Decoupling in the GP16 Dataset

Three main methods were compiled to analyze analyte coupling phenomenon in the EPZT GP16 transect: depth profiles, differences between the concentrations after applying a scale factor (section plots), and direct comparisons of the analytes (concentration ratios as well as spot tests). Depth profiles of cadmium, phosphate, zinc, silicate, and copper were made to begin assessing the analytes for stations 1-17 of the GEOTRACES GP16 cruise (see Supplemental).

Concentrations of elements in the ocean range 15 orders of magnitude (from ~0.5 mol/kg for sodium and chloride to less than 1 fmol/kg for iridium).2 The first consideration for comparing the depth profiles was how to compare the molality of one analyte to another when they have orders of magnitude differences in concentration. A ratio was required to convert units of µmol/kg phosphate to pmol/kg cadmium and nmol/kg zinc to µmol/kg silicate. The new values are of arbitrary units, but they allow for an easier comparison while still representing the way the analyte concentrations change with depth. Options for choosing a ratio to scale by include to set a single global ratio for different water columns16 or to scale by a ratio representing regions such as water masses or oceans.5,9,15,20 For instance, a Cd/Pdeep ratio of 250 (pM/µM) has been reported in literature for the North Atlantic assuming the ratio is relatively constant throughout the deep oceans.16 Janssen et al. opted for two ratios, 0.25 for Atlantic samples and 0.35 for Pacific and Southern Ocean–subantarctic samples.5 Middag et al. reports using Cd:PO4 remineralization ratio of 0.21 ± 0.01 nmol/µmol for the Atlantic deep water values9 but acknowledges that Baars et al. has found ratios with higher Cd content in Antarctic waters, between 0.48-0.65 nmol/µmol.15

In this work, we added an additional approach to determining the scaling factor: the bottom depth concentration ratio between the two analytes for each separate station. This was done in an effort to account for some regional variability while staying well away from the biological activity in the surface ocean. That being said, it should be acknowledged that the bottom depth ratio reflects the source of the bottom water more closely than the water column above it[9] and that with

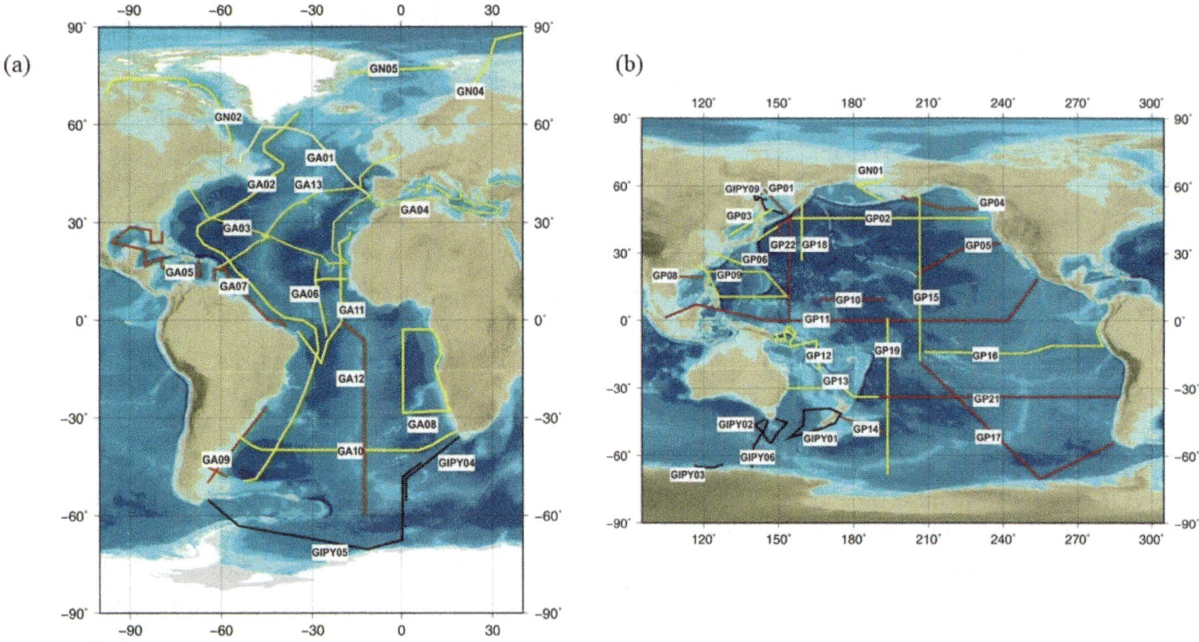

Figure 1. Data from GEOTRACES Pacific cruises GP10 and GP16 (a) were used to investigate methods of assessing coupled analytes27 and (b) Atlantic cruises GA02, GA03, and GA10.28

this approach the deep-water ratio and thus the normalization is extremely sensitive to the depth of deep-water sampling. We use the concentration ratio at the bottom depth as a scale factor to normalize one of the concentration profiles before taking a difference. These difference plots will be compared against plots simply taking the ratio.

The bottom depth ratios from the Atlantic and Pacific cruises GA02, GA03, GA10, GP10 (**Figure 1a and b**), obtained from data in the Intermediate Data Product[3], were each compared to the ratios found at the bottom depths of GP16 stations and found to be reasonably close (**Figure 2a and b**), enough to justify using the bottom depths as scaling factors. The GP16 cruise fell within the values from the other cruises, so a bottom depth ratio may be a way to find a scaling factor while dealing with some of the inconsistencies from one water column to the next.

The data for phosphate and zinc were scaled with their deep-water ratio from that station, noted as Cd/P_{deep} and Si/Zn_{deep} respectively, before being compared to cadmium and silicate. The scaling and differences between analytes were taken as follows: 2

The coupled analytes can be plotted as depth vs analyte concentration on the same graph to see a superimposed comparison of shape between the plots (**Figure 3a and b**). Using the bottom depth concentration values to scale by sets the bottom depth

$$Si^* = Si - \left(\frac{Si}{Zn_{deep}}\right) Zn \qquad \textbf{Equation 1.}$$

$$Cd^* = \left(\frac{Cd}{P_{deep}}\right) P - Cd \qquad \textbf{Equation 2.}$$

difference at zero. Any deviation away from zero represents greater decoupling, so the analytes are the most decoupled at the depth of the maximum deviation from zero. To investigate how decoupling maxima relate to oxygen minima, plots of the oxygen concentrations vs depths were created and superimposed onto the difference vs depth plots (**Figure 3c and d**).

Plotting the slope of the Cd* and Si* plots vs depth is also shown (**Figure 3e and f**). This is in essence a discrete optimization problem, looking for optima by locating where the derivate plot crosses the domain axis. A moving two-point window average of the slope was one attempt to identify the maximum peak (represented by a slope of zero) by reducing noise. Two points were chosen as more points trades off a decrease in noise

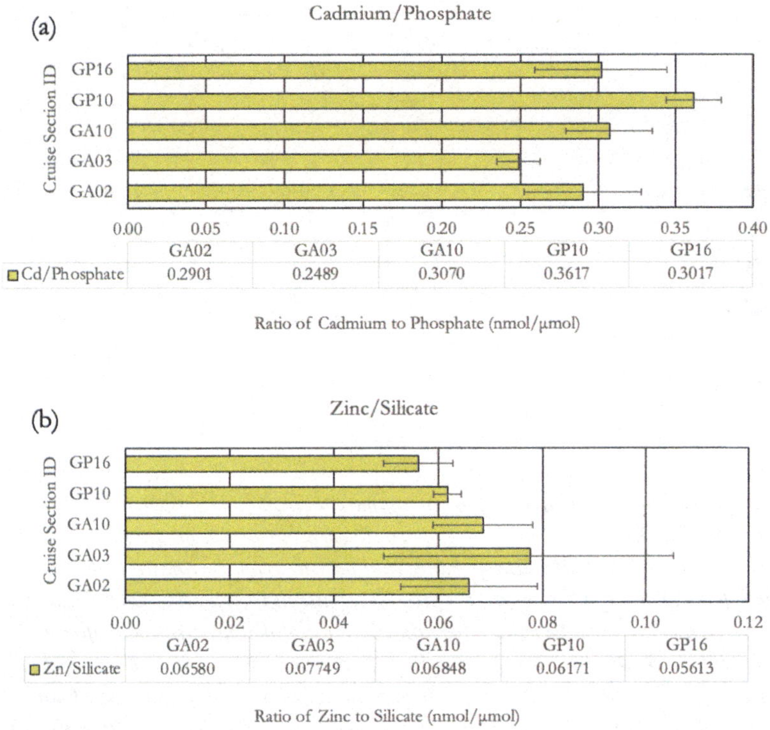

Figure 2. Comparisons between bottom depth concentration ratios found throughout separate GEOTRACES cruise sections for cadmium to phosphate (a) and zinc and silicate (b). Errors Bars represent ± 1σ.

for a decrease in sensitivity. Because it graphs the slope at each point, the first and last point are not accurate, but the points of interest are well represented in the upper region of the water column. Because the upper water column is the main focus, only the upper 1500 m of the water column is shown. Each decoupling maximum was compared to regions of oxygen minima in **Table 1**.

Section Plots

Ocean Data View (ODV) was used to create weighted average gridding section plots of the depth profiles for 17 stations of the GP16 transect (**Figure 4**).

Spot Test

The spot test simply takes a ratio of concentrations at a single point. It does not require that the transect be broken up into water masses beforehand as was done for here (**Table 2**) and can provide useful information from a single sample. If the spot test is needed to represent a wider region, multiple spot tests can be averaged and reported along with their standard deviation. Alternatively, many spot tests could be conducted for a single point over time and averaged to get a sense of how the ratio changes over time.[15] To assign spot test ratios to a given water mass, the optimum multiparameter analysis (OMPA) performed by Peters *et al.* for the GP16

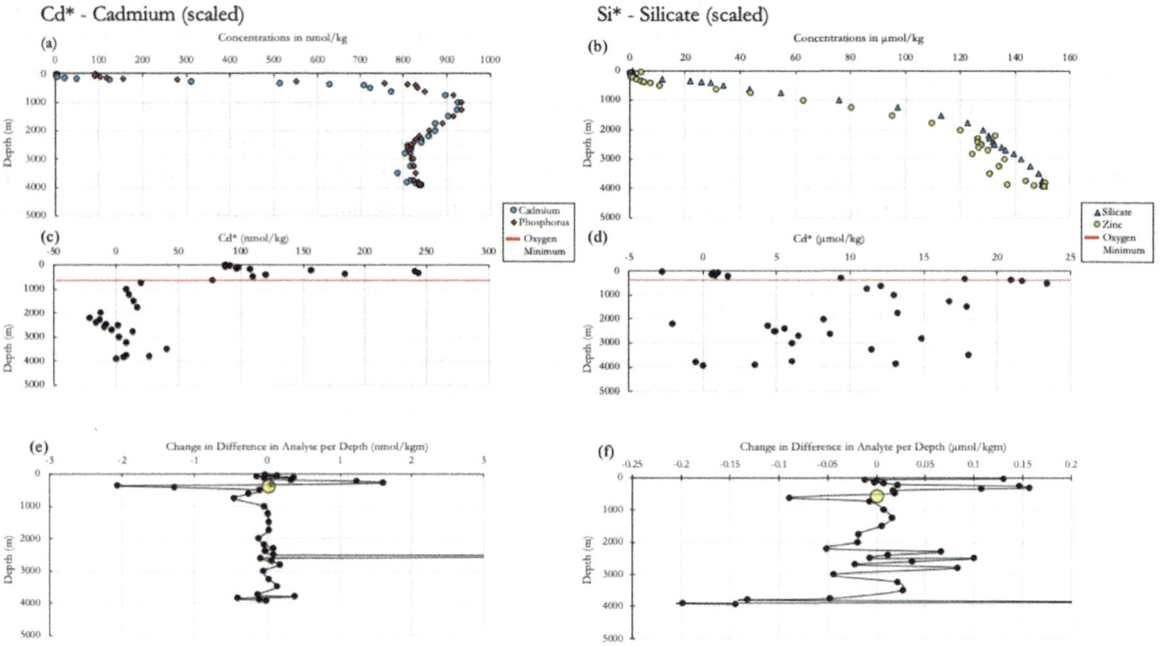

Figure 3. GP16 Station 15 data was used to plot coupled analytes to compare their shapes by superimposing their depth profiles after scaling with the deep-water ratio (a) and (b). Differences between these two curves allow for the greatest deviation between coupled analytes to be found as the greatest difference in (c) and (d). The rate of change in difference is approximated by plotting slope vs depth between points from the difference vs depth plot (e) and (f). Lower depths provide less useful information when looking at the change in differences, but in the upper 1500 m (our region of interest) the point where the change in differences reaches 0 after a local maximum in change correlates closely to the maximum difference: the maxima of decoupling of the two analytes (circles in (e) and (f)). The point of maximum decoupling is circled in plots (e) and (f). The region of maximum decoupling is where the change reaches 0 after reaching the maximum rate of change. This can be interpreted as the entire interval of depths between the last positive of a maximum and the first negative rate of change.

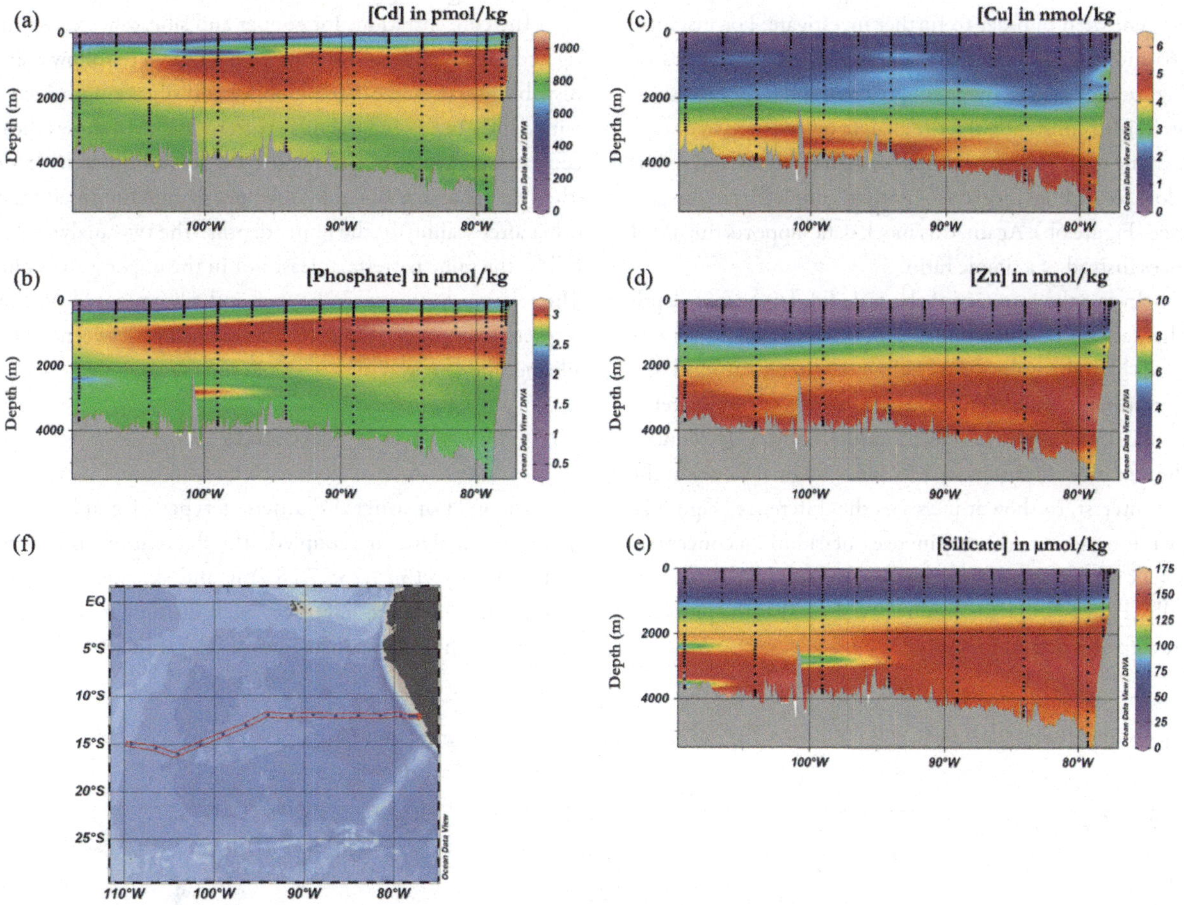

Figure 4. Stations 1-17 along the GP16 transect, and corresponding section plots of the micro and macronutrients investigated in this study. Station 1 is the most inshore while 17 was the furthest offshore station of the transect.

cruise was used to help select the locations for sample ratios.[1] An OMPA uses computer modeling to determine the source origin for different water masses within a region.

Guidelines for Oxygen Minimum Zone (OMZ)

The OMZ was defined using the limits set by the IOC and UNESCO in a summary to policy makers regarding declining oxygen levels in the ocean: oxygen poor regions were defined as anoxic with sublethal biological effects at < 63 µmol/L of O2 and as an OMZ at < 20 µmol/L.8 Ranges were tabulated along with the maxima of the analyte differences for comparison (Table 1). The table only contains maxima in the region of interest within the upper water column since there were occasional deviations much further down the water column.

Initial Analysis

Determining the impact of using differences or using a ratio to compare analytes was done with the help of plots of the real data as well as mock data. These initial plots were very useful in seeing how coupling presented itself in different plots and how to best represent it. Basing the comparison off of differences gives better detail and a slightly different maximum than basing the comparison off of ratios (**Figure 5**). For this reason, the final analysis (**Figure 9**) was conducted as the difference between scaled values.

Mock Data

It is important to note the difference between the two because, for a concentration ratio to remain constant along a water column, the differences between the analyte's concentrations would have to change accordingly to keep up. Mock data was generated in order to further investigate. For instance, Figure 6a shows analytes that are coupled with a constant ratio, while Figure 6b shows analytes coupled that differ by a constant amount. With a constant ratio, the analyte represented in orange has to assume more drastic net changes than the blue, and so it is not as close to the blue's curve as it is when coupled by a constant difference (Figure 6b). Again, this mock data supports the use of differences instead of a simple ratio.

To help test the integrity of the scaled differences vs depth plots, the data on the first 1000 m for stations 14 and 15 was taken and had the values for cadmium altered around the oxygen minimum zone. Figure 7a and b shows the unaltered version, and then how a loss of cadmium, arbitrarily reduced the cadmium concentrations by 1.5x (66% original value) in the region of interest, to show it increases the difference (Figure 7c and d) while a gain, or arbitrary increase of cadmium concentration by 1.1x in the region of interest, can decrease the difference drastically (Figure 7e and f). This helped show that the plots did visualize a greater decoupling.

Excluding Copper and Silicate

The concentrations for copper and silicate were also investigated for coupling, as they have been shown to follow a similar distribution as each other.2 Copper and silicate both increased with depth, but not in a similar pattern. Figure 8 shows the concentrations of silicate and copper for several different stations. The two values are not able to be neatly superimposed on each other after scaling for the upper depths. The two analytes do not follow the same pattern, at least not in the upper water column. The relationship between copper and silicate could not be interpreted the same way as the relationship between zinc and silicate, and so it was omitted from further analysis.

Results

The goal of using the different types of graphs is to tell if the two analytes are coupled. The fluctuations of cadmium and phosphate,15,21,23,27,28 zinc and silicate,13,22,23 and copper and silicate18 have been well documented as coupled in their oceanographic distribution. The ability for the various

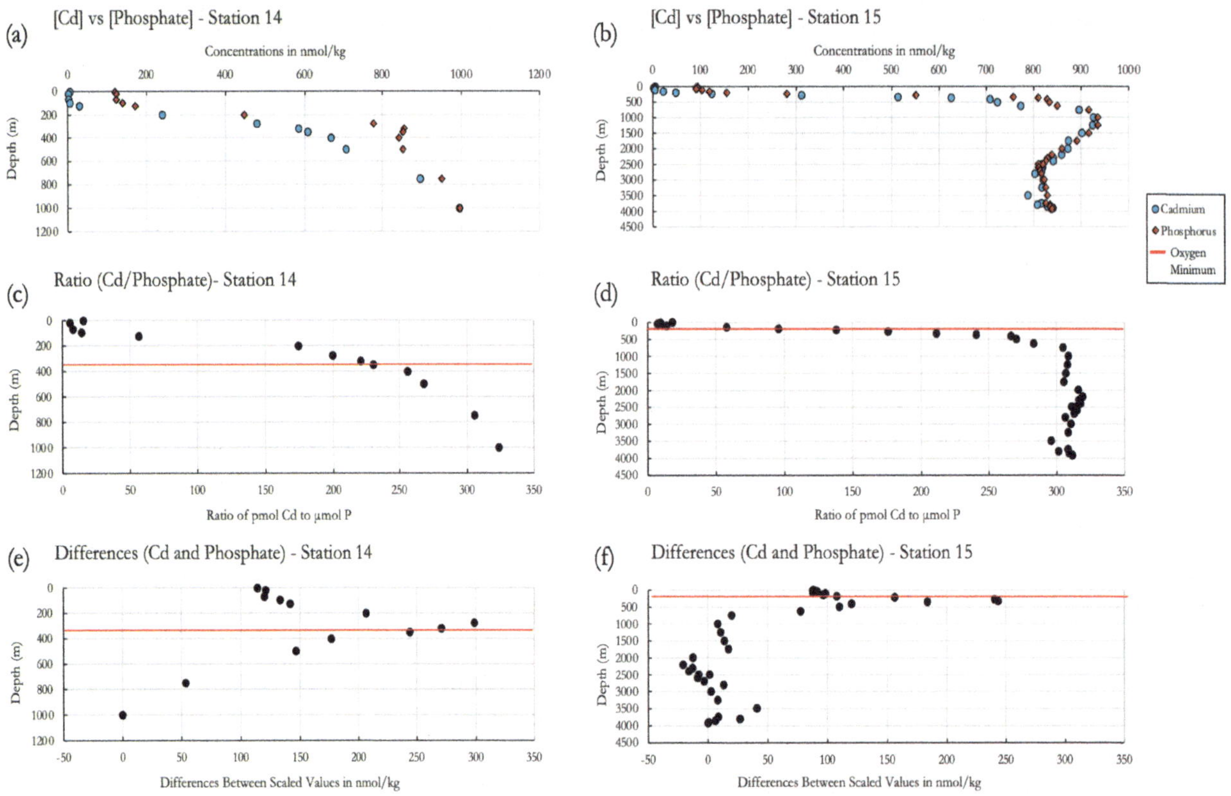

Figure 5. The greater difference in the superimposed values is shown well in (e and f) which compares using differences, but not as well when comparing by ratios (c and d).

methods to evaluate coupled trends was gauged using data from the GEOTRACES EPZT GP16 cruise. We were able to detect the decoupling for cadmium-phosphate and zinc-silicate along the Peru transect using the section plots of Cd* and Si* against depth and longitude as well as with the ratio plot.

Depth of Max Based on Scaled Differences

A bottom depth scaling value allowed for an initial analysis of the differences in analyte concentrations. To help determine if the decoupling correlates with oxygen deficiency, the IOC-UNESCO definitions for hypoxia (< 63 µmol/L of O2) and OMZ ranges (< 20 µmol/L of O2) acted as guidelines to quantify oxygen deficient waters within our stations.8 Table 1 lists the results for stations 1-17 of GP16. The maximum differences are recorded as the single depth where the sample with the maximum difference in concentration was collected. Dissolved oxygen (DO) for that point is also included. The oxygen minima are recorded as the range of depths that were either hypoxic or within an OMZ range.

Cd* and Si* vs Depth

To better understand and visualize where the decoupling is occurring, the graphs relating Si* and Cd* to depth for each station (as in Figure 2b and f) were translated into ODV plots (Figure 9). Plotting against longitudes (Figure 9a and b) helps preserve information about what regions exhibit analyte deviations from coupling. The decoupling pattern becomes more easily identified as the bright red regions. While the OMZ seen

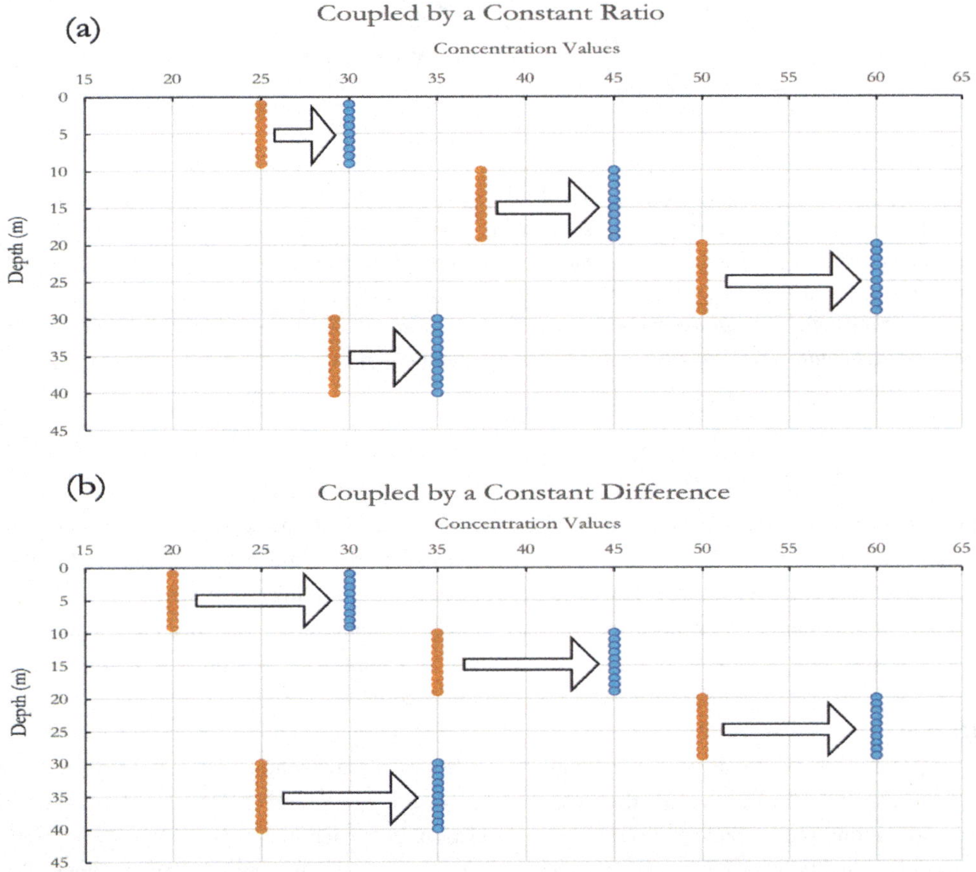

Figure 6. The orange and blue represent theoretical concentrations of different analytes changing with depth, where the analytes are coupled either by a constant ratio (a) or a constant difference (b). Coupling by a constant difference (b) produces more similar rates of change to those observed. Note that the 25m depth arrows are the same distance in both panels.

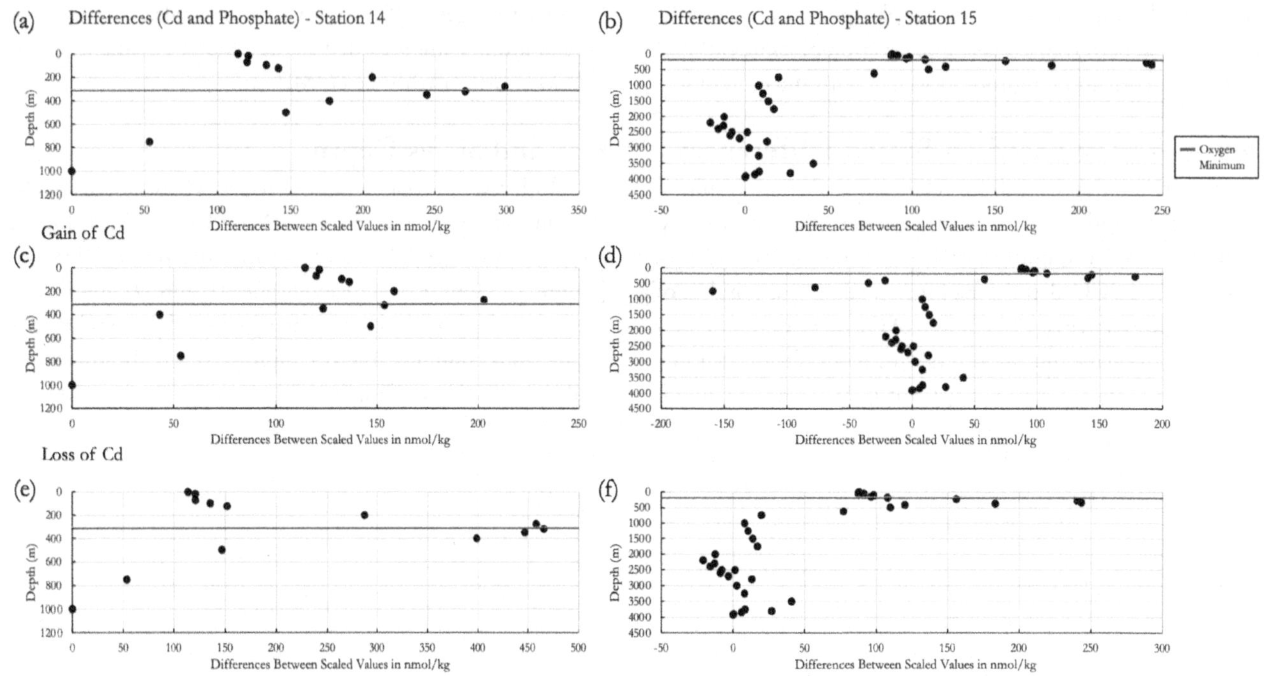

Figure 7. As a precautionary check, mock data was used to depict how the graph of differences vs depth can be altered by an increase and decrease of cadmium within the range of interest. Differences between cadmium and phosphorus are shown and give profiles for station 14 and 15. Unaltered data (a and b) is shown for the two stations, as well as the change when there is a gain of Cd (c and d), and the change due to losing Cd (e and f) within the region of interest. This helped show how using scaled differences can visualize the depth of the decoupling, which correlates as expected with the oxygen minimum (red line). Gain of Cd corresponds to an increase to 110% cadmium concentration. Loss corresponds to cadmium concentration of 66% the original (arbitrary).

in Figure 9c dwindles with increasing oxygen concentrations to the west, the plots of Si* and Cd* to depth do not exhibit the same decrease. This suggests that the decoupling event is not directly related to an oxygen minimum as previously literature has suggested.

Concentration Ratios

Unlike comparing Cd* and Si* to depths and longitudes, plotting two concentrations against each other does not preserve information about the regions of decoupling. However, this approach is valuable in investigating how the concentrations relate over an entire transect, and the loss of locations can be partly amended by overlaying a third axis representing the positions of samples (e.g., Figure 9d and e). Using this plot, coupled analytes display a linear relationship between their concentrations along a water column. A bilinear plot, such as those shown and highlighted in Figure 9d and e, represent a decoupling event9 where the analytes were coupled, underwent some event that decoupled them, and then once the external influence was no longer applied, they returned to a linear coupled state.

For phosphate and cadmium, the greater differences were in the upper water column where the oxygen concentration was minimal (Figure 9a and c). Zinc and silicate exhibit a similar correlation, but much looser. Notably, oxygen levels in the upper water column change over the transect due to Peru having one of the largest OMZs right offshore.11 This is corroborated by the purple oxygen minimum region in Figure 9c that shows the OMZ range of depths narrowing significantly once further offshore than approximately 100 degrees West.

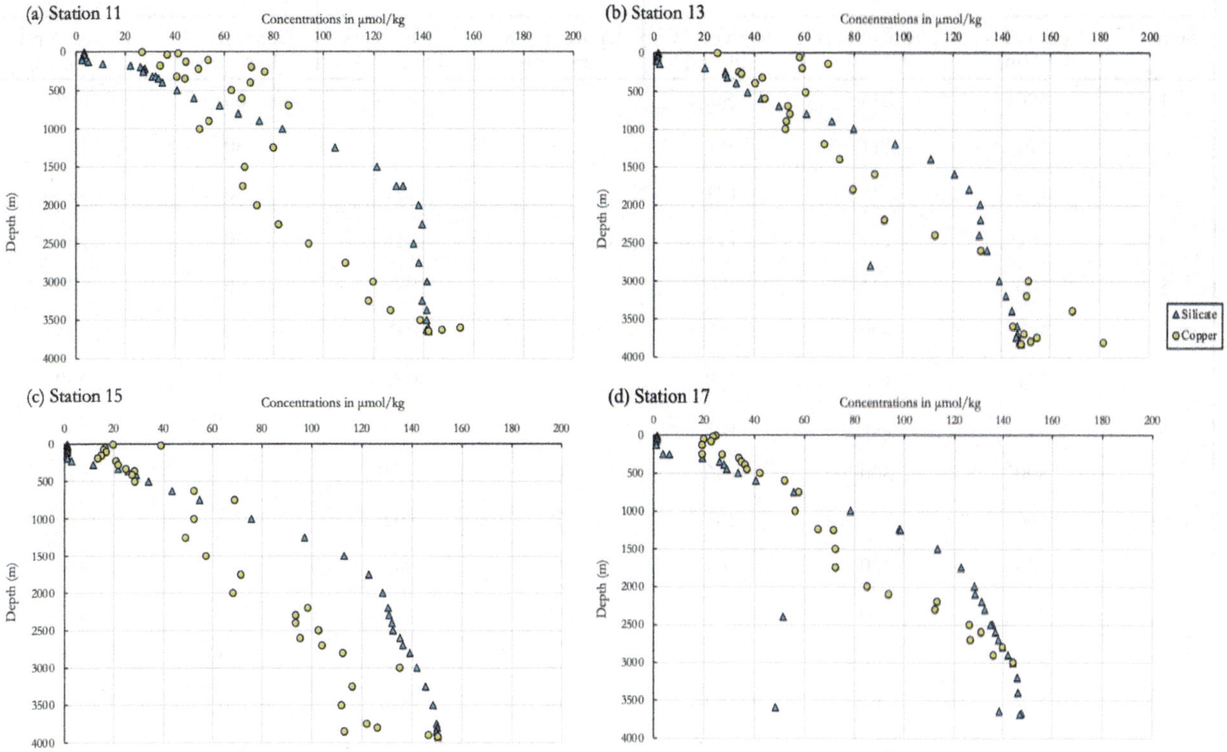

Figure 8. Copper and silicate are compared where the lowest point is set as the scaling factor. The two analytes do not follow a similar curve, especially not in the upper waters where the oxygen minimum occurs (a-d).

Spot Ratios

Similar to the concentration ratios but at a point in space, a spot ratio can represent the concentration ratio for a given water mass. It does not necessarily show decoupling of analytes alone, but it is useful in conjunction with the aforementioned graphs for a clearer picture. The GP16 cruise data product had to be split up so that a sample from each water mass could be used. Peters et al.1 used an OMPA to determine the percentage contributions of different water masses. The Equatorial Subsurface Water (ESSW) and Pacific Deep Water (PDW) showed relatively less mixing making them more representative of a single water mass and were used to find a sample that would adequately represent just that water mass. The other regions had water masses that were more mixed, and they were reported as their constituent water masses (Table 2). A single ratio helps to quickly compare spatially disjunct regions and allows researchers to quickly compare whether two regions are similar. This can be used to tell whether a dynamically interesting region like the phototrophic zone's concentration ratios match that of what we expect from a more stable zone such as further down the water column.

Discussion

Standardizing a Scale Factor

The goal of this study was to compare the patterns of change between two analytes at very different concentrations. To compare a trend, plots of concentration vs. depth were made for each analyte, and a scaling factor was used to compare their depth profiles. A difference between the concentration of one analyte and the scaled concentration of another (e.g. Figure 5e and f) provides a means to assess where the analytes maintain a coupled relationship. In this work, bottom depth ratio between analytes was set as the scaling factor, which in turn tethers the zero value to the bottom depth the ratio was measured. Any deviation from zero can be considered a location with decoupling. If a global average is used instead to scale the concentration value of one analyte, then the values will be centered around some non-zero value determined by that global ratio. Deviations from this value are analogous to deviations from zero in the bottom depth scale factor plot.

			Oxegyn Levels	< 63 µM		< 20 µM	
Station	Depth of Max Cd* (m)	Depth of Max Si* (m)	Depth of DO Min (m)	Upper Depth Limit (m)	Lower Depth Limit (m)	Upper Depth Limit (m)	Lower Depth Limit (m)
1	300	235	80	66	1000	80	676
2	111	111	90	20	111	30	111
3	171	50	171	20	171	30	171
4	201	111	111	50	755	50	755
5	500	331	380	56	999	56	600
6	301	400	95	81	1000	95	601
7	251	126	150	110	1000	110	750
8	280	227	161	126	999	126	451
9	300	300	136	120	1000	120	625
10	166	166	201	166	802	166	399
11	236	320	201	180	901	180	401
12	230	230	230	211	1000	211	401
13	250	250	250	201	1000	250	400
14	275	401	320	275	1001	320	350
15	278	500	361	332	750	361	-
16	295	351	351	351	601	351	-
17	245	351	351	301	500	351	386

Table 1. Comparing Dissolved Oxygen (DO) Minima to the Maxima of Decoupling.

These plots provide an abundance of specific data on locations where the paired analytes diverge.

The biggest fault of these graphs is the difficulty in selecting a scaling factor that is appropriate. It would be beneficial to find a globally applicable method to determine a scale factor to allow researchers to compare concentration profiles between locations, however it might not be possible to find a globally relevant value: Quay et al. has reported that interbasin variation of particulate Cd/P exported from the surface and subsequently degraded indicate that a consistent bottom Cd/P slope is not expected and that a ubiquitous ratio is unwarranted21. Figure 2a shows broadscale agreement in deep-water Cd/P and Si/Zn across ocean basins, and perhaps one average value could be selected not because it is perfect, but because having an agreed upon scaling factor could be a useful tool.21

There are also some issues presented by using a bottom depth value as a scaling factor that were found and must be acknowledged: the sensitivity to sampling of that bottom depth sample the ratio is determined from the variation in bottom depths along the transect, and that the bottom depth represents a ratio from the region that water mass comes from and not the water column above it.9 There is also the issue that copper analysis showed, likely from being more heavily impacted by processes such as co-scavenging4 and remineralization from the sea floor7 which skew the bottom depth ratio. Though using a bottom depth ratio had its faults and difficulties, it was able to consider specific station variation without instilling bias by using another author's global average ratio of coupled analytes (Table 1). One possible alternative to this method is to include a normalization ratio developed from pre-formed signatures for each water mass instead of the bottom depth water mass. Averaging concentration ratios over entire water masses would have to be approached with careful consideration but may provide a better scaling factor. It would aim to give the improved reproducibility between datasets that a global average gave while also representing some local variation that was captured by the bottom depth value. Pre-formed concentration ratio signatures are used in literature in the analysis of isotopes of iron,2,18 isotopes of lead,29 isotopes of cadmium,5,16,30

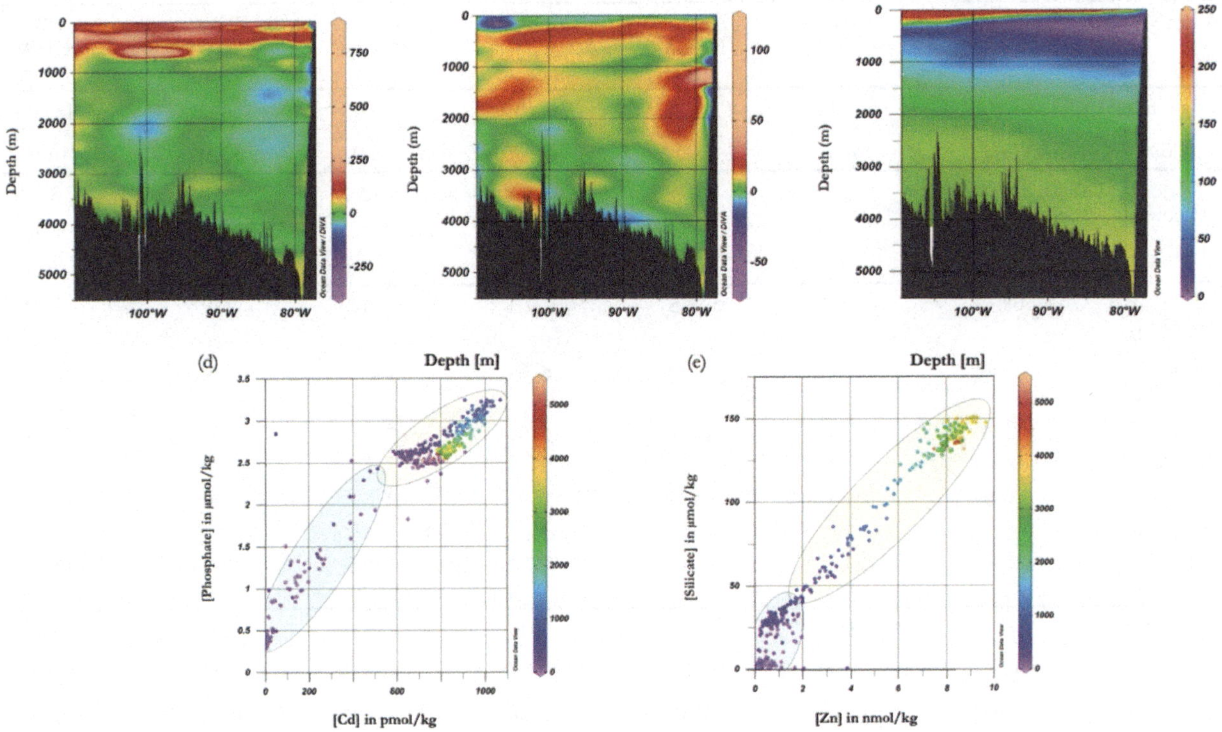

Figure 9. The extent of decoupling is shown over the transect with greater decoupling represented by a larger difference between pairs phosphate and cadmium (a) as well as silicate and zinc (b). This shows as a red band in the upper waters. Note the OMZ distinguished as a purple band of low oxygen (c). Both the decoupling event (shown in a and b), and the OMZ (shown in c) occupy similar depths, but the decoupling event extends further west than the OMZ region without reduction in intensity. The decoupling is shown again for phosphate and cadmium (d) as well as silicate and zinc (e), with the bilinear nature of the plots - indicating the decoupling – highlighted with colored ellipses. Each ellipse represents the nutrient profile under two different stable regimes.

isotopes of zinc,[16,22] and isotopes of helium.[1] Some pre-formed concentration ratios are also reported between two unrelated chemical species including ratios between cadmium and phosphate,[9,15,23,31] zinc and silicate,[13] and for zinc/silicate/cadmium,[23] but there is little consistency between reported methods that incorporate preformed signature for specific water masses. Also, though preformed signatures may address several limitations associated with using the bottom depth ratio, issues arise in the complexity of having to annotate each point with its water mass in order to consider a different nutrient coupling ratio for each water mass. Accuracy and simplicity of analysis must be balanced to find methods to scale the concentration profiles by so that deviations in coupled analytes can be measured more precisely and be compared between researchers.

Direct Concentration Ratios

Coupling can be determined by plotting concentration of one analyte to another. A plot directly comparing two concentrations is less time intensive than scaling and taking a difference, and it can be created from a GEOTRACES data product with significantly less effort. A constant ratio between analyte concentrations conveys that they remain proportional with a linear trend, and, inversely, a linear regression can be used to determine an expected concentration of one analyte using known concentrations of another element coupled to it. A bilinear trend means that the coupled analytes underwent a decoupling event, and then once that event passed, they had returned to being coupled. An R2 value can be used to quantify how linear the relationship is,[18] or such plots can be used to see how the profile adheres to a conceptual conservative mixing line.[19]

Figure 6 highlights that coupling is sensitive to how it is defined. A completely linear concentration ratio implies the difference between values should grow or shrink proportionally. A linear concentration ratio will lead to varying scaled differences; and therefore, consideration should be given that the linearity

Water Mass	ESSW	ESSW/ SPCW/ ESPIW	AAIW/ EpPIW	PDW	PDW/ UCDW	PDW/ LCDW	AABW/ LCDW
Longitude	-81.50	-115.00	-99.00	-84.00	-99.00	-84.00	-152.00
Latitude	-12.00	-14.77	-14.00	-12.00	-14.00	-12.00	-10.50
Station Number	6	21	13	7	13	7	36
Depth (m)	300.9	2200.3	900.2	2751.3	1800.3	3625.2	4000.1
Cadmium pmol/kg	608.3	805.1	983.5	872.1	934.3	789.5	735.9
Phosphate µmol/kg	2.620	2.530	3.110	2.760	2.830	2.640	2.420
Zinc nmol/kg	0.7315	8.138	3.492	8.701	7.259	8.065	7.353
Copper nmol/kg	0.6059	2.983	1.476	3.593	2.227	4.471	3.480
Silica µmol/kg	31.23	123.7	71.16	147.0	126.5	138.4	133.1
Cd:P	232	318	316	316	330	299	304
Zn:Si	0.0234	0.0658	0.0491	0.0592	0.0574	0.0583	0.0552
Cu:Si	0.0194	0.0241	0.0207	0.0244	0.0176	0.0323	0.0261

Table 2. Analyte Ratios at Given Points; ESSW = Equatorial Subsurface Water; ESPIW = Eastern South Pacific Intermediate Water; SPCW = South Pacific Central Water; AAIW = Antarctic Intermediate Water; EqPIW = Equatorial Pacific Intermediate Water; PDW = Pacific Deep Water; AABW = Antarctic Bottom Water; LCDW = Lower Circumpolar Deep Water; UCDW = Upper Circumpolar Deep Water

of concentration ratio plots may not represent the strongest correlation index. A constant scaled difference provides a more accurate definition to show when two analytes are not aligned or superimposable in their depth profiles.

The Spot Test

The spot test is very similar to comparing the ratio of concentrations of analytes. The spot test takes a concentration ratio or an average concentration ratio within a water mass to compare it to different water masses. The spot test simplifies portions of the data into a single reportable value or an easy-to-read table. A spot ratio is simpler to make and interpret than a graph comparing the concentration ratios. While Table 1 characterizes a huge area of ocean with just a handful of ratios, it does not convey as much detail on specific regions like a depth profile or transect plot would. The spot test can represent a water mass with less requirements, but it can be elaborated on further such as graphing spot test values over time for a given region.32

A plot over a region considers the range of possible concentrations while the spot test represents the actual values of the analytes at a spot, and for that reason, the spot test's interpretation is not as biased by other points.9

Using Depth Profiles

Depth profiles are time consuming to make because a new plot must be created and interpreted for each station and for each analyte. It was useful to have depth profiles for our transect when a specific region needed to be analyzed further, but so many can distract from the key region and become a needlessly exhaustive effort.

Plotting the concentration vs. depth for each station makes it easier to extract values for a given station and highlight the depths decoupling occurs when two plots are superimposed on each other. It falls behind the other methods in that it displays such a small portion of any cruise's data product that more graphs are needed to represent data from the transect. A less time-consuming method would be to start with an ODV plot of the entire transect before investigating specific stations to further investigate a specific water column. An ODV plot of the entire transect contains essentially all the information several depth profiles would, with the caveat that certain values may be harder to discern directly off an ODV plot.

Limitations to Consider for Nutrient Coupling Ohnemus et al. used factor analysis to assess relationships between analytes without the need for any ratios or normalization.4 They also chose to omit spatial information (such as station number, latitude/longitude, and depth) or other chemical variables (dissolved oxygen/macronutrient concentrations) in the analyses, because they suggest these variables are likely to affect particulate phase concentrations in nonadditive or nonlinear fashions. Their analysis resulted in many elements being categorized into 6 types based on their depth profiles. They focus explicitly on particulate data and well-defined basins that have strong particle locality due to the persistent downward settling of particles. They suggest this allows 3d dissolution (which more greatly affects dissolved concentrations as they settle) to be ignored partly in the way data are taken. Factor analysis can quantify variables latent or difficult to directly measure variables that are still thought to drive the variability of measured variables and elucidate potential shared transport mechanisms that affect multiple elements. In this case, particulate trace element concentrations (the measured variables) are presumed to be the result of additive particulate phase abundances (the unmeasured or latent variables), even if the individual particulate phases themselves are difficult to separate analytically. Ohnemus et al. reported that they were able to perform the statistical analysis independent of spatial information such as station number, latitude/longitude, and depth or other chemical variables like dissolved oxygen or macronutrient concentrations in the analyses, ideal features to an improved analysis as these variables likely affect particulate phase concentrations in nonadditive or nonlinear fashions. This method has only had success with certain analytes so far, but statistical methods such as these that avoid bias by not requiring as much initial input are ideal to standardizing analysis.

The limitations seen with comparing copper concentrations highlight that analytes coupled in distribution are the exception, and that comparisons of different analyte pairs must be on a careful case-by-case basis. Though copper is incorporated into biomass, copper concentrations produce different depth profiles that could not be superimposed regardless of the scale factor used (Figure 8). Copper's concentration profile increased relatively linearly with depth, in contrast to the other analyte profiles which had a sharply diminishing increase in concentration with depth. Copper is influenced heavily by rapid surface water scavenging as well as in-situ deep water scavenging and bottom water injections that recycle copper back into the overlying water column, and it does not appear to show a nutrient like distribution in the deep ocean waters.7 Both silicate and copper concentrations have increasing concentrations with depths, but they do not show a similar curve when superimposed.5 Jacquot & Moffett corroborated reports that copper and silicate do not share a strong correlation in their distribution in the water column above 2000m.18 The lack of correlation may be due to the exceptionally deep maximum for silicate and a slowly increasing but stable rate for copper concentrations below 1000m. Interestingly, Zn is also known to have significant hydrothermal inputs from the sea floor within the GP16 transect,20 but did not exhibit a similar distribution pattern to copper. Roshan suggests that Si has a heavy input from river systems and Zn from hydrothermal vents, but the biological control imposed on both of them in the upper biological active region by the uptake of nutrients and subsequent sink has such profound effects that it couples them over vast areas of the ocean. The factor analysis by Ohnemus et al. shows that copper is more similar to elements scavenged throughout the water column (similar to iron and manganese hydroxides that settle), and that zinc belongs grouped as a nutrient type.4 The stark contrast between the two may be largely attributed to persistent iron oxide scavenging of copper as it settles in the water column. Factor analysis helps justify the lack of similar profiles between copper and silicate in the upper water column, but it does not give enough information in terms of where or if the analytes decouple (only an index of similarity between the two factors' transport mechanisms).

Peru results (Oxygen and Sulfur's Role) The plots made (Figure 9) suggest a correlation between depths of oxygen minima and the decoupling of cadmium-phosphate along with the zinc-silicate. The correlation between the oxygen minimum and decoupling of nutrients may be due to the depth of the water column which relates them indirectly through biological uptake in the euphotic zone. Figure 9a, b, and c highlight that while oxygen minima and the greatest decoupling are found in approximately the same depths, the decoupling is still observed further offshore outside the oxygen minimum zone. This is likely due to other factors than solely CdS precipitation, as CdS was not found in extraordinary amounts along this transect.16 Bourne et al. already has elaborated on six specific factors driving the variation observed in particle concentrations in regards to cadmium cycling: (1) passive uptake by marine organisms, (2) deliberate uptake (in the case of Cd and diatoms as a substitute in low Zn culture), (3) bio-dilution (biological ratios can change if constant uptake of a trace mineral is met with changing growth rates), (4) species composition (especially cyanobacteria vs microscopic eukaryotes), (5) cadmium sulfide (CdS) precipitation in euxinic microenvironments of sinking aggregate particles, and (6) the uptake is proportional to dissolved concentrations (and the dissolved concentrations change a lot - the author specifically points to variation due to latitude).31 Their results support 1-3: iron limited conditions yield higher Cd:P ratios (3) and that low concentrations of dissolved Fe, Zn and Mn can lead to increased uptake of Cd (1 and 2). Bourne also found the variation noticed in the Pacific to be unlikely caused by the oxygen minimum or sulfur leading to cadmium precipitation. Other authors have suggested that the loss of dissolved cadmium is due to preferential uptake by microbes,14 but more attention to the different ways a microbe can interact with biogeochemical cycles is needed. Biogeochemical cycling is very sensitive to species composition, and special attention is required to the forms of life found so that the key interactions previously mentioned can be better characterized for an area. The differences between organisms such as prokaryotic cyanobacteria and eukaryotic diatoms are so great, that their roles in biogeochemistry cannot be oversimplified if an accurate representation is trying to be achieved.4 Biochemical flux should be considered from the perspective of multiple biological pools, and deeper investigation of all the complex biological interactions is required before the necessary biological factors are delimited.

The reason for decoupling most likely begins with biota. Some specific ways that microbes may interact with settling elements in this transect are presented herein for the sake of drawing analogy with other analytes. Microbes have been posited to be present in higher amounts within the euphotic zone and subtending OMZ which causes a biogenic sink of both Cd5,18,21,23,31 and Zn.13,16,20,22 Some authors suggest that these microbes increase the sulfur levels as part of a cryptic sulfur cycle leading to CdS formation (though it is unclear whether this is the major driving force).10 Another cause of biogenic sink may be due to the uptake of Cd due to organic ligands and a subsequent sink in the concentration of cadmium.27 Similarly, organic ligands cause a sink for many elements such as Cu.18 Bourne et al. mentions that P lability changes depending on what biochemical it constitutes.31 They considered phosphorus bound in ATP or DNA to be more labile than some other pools such as phospholipids, and they found Cd to have a lability between those two. This highlights the significance of determining what molecules the element gets biochemically sequestered into has on remineralization. Ohnemus et al. suggests that the initial aggregation and export of surface particles and especially the subsequent biological accumulation that occurs right below the surface drives concentration decreases as they settle and scavenge.4 As the particles sink, the influence of biological drains becomes less intense and particle disaggregation begins to dominate. Finer refractory particles tend to accumulate. Phosphorus and cadmium have similar depth profiles due to a lack of external input and the influence of the above dynamics. It seems that each analyte has an independent depth profile, and that the major groupings (such as a nutrient-type curve) are only able to suggest what the major driving force is.

Conclusion

This paper sought to review approaches for investigating nutrient coupling and to suggest a new approach using the data from the GEOTRACES GP16 cruise to be able to characterize the extent of decoupling between solutes with similar distribution types along a transect of the eastern South Pacific. Nutrient type distributions are coupled when solutes in the ocean are taken up and re-mineralized at similar biological rates. Investigating where perturbations from that coupled state occur in the water column provides insight into obscure phenomena that contribute to concentration fluxes which may allow for more realistic modeling of biogeochemical cycles. Here we consider the potential benefits of standardization across methods of assessing nutrient coupling by specifically comparing and contrasting current methods reported in literature.

A more insightful analysis to the correlations between coupled nutrients was achievable by compiling three standard approaches 1) depth profiles, 2) differences between the concentrations after applying a scale factor, and 3) direct comparisons of the analytes. Differences between concentrations after applying a scale factor offered the most insight, though it required a considerable amount more manipulation of the data compared to other methods. The inconsistent usage of scaling factors throughout literature is a challenge for this method; selecting some form of standardization would make this method more readily applicable for comparisons between expeditions and datasets. Basing the comparison off of differences gives better detail and a slightly different maximum than basing the comparison off of ratios (Figure 6).

Plotting the differences between concentrations after applying a scale factor allowed a comparison between cadmium and phosphate to oxygen minima to show that the decoupling extends past the oxygen minimum zone that resides off the coastline of Peru. Because the goal of studying this phenomenon is to learn about the underlying causes of these biogeochemical cycles, being able to characterize the decoupling in a way that is easy to visualize is extremely important to the efficacy of research. Other methods that are quicker occasionally miss the nuance that is required to identify biogeochemical processes from data products.

More work is needed to resolve our understanding of nutrient coupling throughout Earth's systems. The success of these large-scale data collection programs such as GEOTRACES goes hand-in-hand with our increasing ability to interpret data in a standardized way and report the results in a detailed yet clear manner. As the community continues to get more adroit with large datasets, the amount of information we can glean from them will expand as well.

Acknowledgements

The authors thank David Janssen for the generous help he gave on a draft of this work. His feedback helped to shape this paper and push it in the right direction. The authors also thank the chief scientists of the cruise, and ODF for measuring the nutrients.

References

Peters, B.D., Jenkins, W.J., Swift, J.H., German, C.R., Moffett, J.W., Cutter, G.A., et al. (2018) Water mass analysis of the 2013 US GEOTRACES eastern Pacific zonal transect (GP16), Marine Chemistry 201, 6–19. https://doi.org/10.1016/j.marchem.2017.09.007

Bruland K.W., Middag R. and Lohan M.C. (2014) Controls of Trace Metals in Seawater, Treatise on Geochemistry, 2nd ed., 8, 19–51. https://doi.org/10.1016/B978-0-08-095975-7.00602-1

Schlitzer, R., Chemical Geology (2018a) https://doi.org/10.1016/j.chemgeo.2018.05.040

Ohnemus, D. C., Torrie, R., & Twining, B. S. (2019) Exposing the distributions and elemental associations of scavenged particulate phases in the ocean using basin-scale multi-element data sets, Global Biogeochemical Cycles, 33, 725–748. https://doi.org/10.1029/2018GB006145

Janssen, D.J., Conway, T.M., John, S.G., Christian, J.R., Kramer, D.I., Pedersen, T.F., Cullen, J.T. (2014) Undocumented water column sink for cadmium in open ocean oxygen-deficient zones, Proceedings of the National Academy of Sciences of the United States of America, 111(19), 6888–6893. https://doi.org/10.1073/pnas.1402388111

Williams, G.R. (1987) The coupling of biogeochemical cycles of nutrients, Biogeochemistry, 4(1), 61–75. https://doi.org/10.1007/BF02187362

Bruland, K.W. (1980) Oceanographic Distributions of Cadmium, Zinc, Nickel, and Copper in the North Pacific, *Earth and Planetary Science Letters,* 47, 176–198. *https://doi.org/10.1016/0016-7037(81)90099-5*

Breitburg, D., Grégoire, M., Isensee, K., Chavez, F., Conley, D., Garçon, V., *et al.* (2018). The Ocean is Losing its Breath: Declining Oxygen in the World's Ocean and Coastal Waters, *Intergovernmental Oceanographic Commission of UNESCO (IOC-UNESCO),* 10.

Middag, R., van Hueven, S.M.A.C., Bruland, K. W., and de Baar, H.J.W. (2018) The relationship between cadmium and phosphate in the Atlantic Ocean unraveled, *Earth and Planetary Science Letters,* 492, 79–88. *https://doi.org/10.1016/j.epsl.2018.03.046*

Gilly, W.F., Bernan J.M., Litvin, S.Y., Robsion, B.H. (2013) Oceanographic and Biological Effects of Shoaling of the Oxygen Minimum Zone, *Annual Review of Marine Science,* 5, 393–420. *https://doi.org/10.1146/annurev-marine-120710-100849*

Paulmier, A., Ruiz–Pino, D. (2009) Oxygen minimum zones (OMZs) in the modern ocean, *Progress in Oceanography,* 80, 113–128. *https://doi.org/10.1016/j.pocean.2008.08.001*

Graco, M.L., Purca, S., Dewitte, B., Castro, C.G., Morón, O., Ledesma, J., *et al.* (2016) The OMZ and nutrient features as a signature of interannual and low-frequency variability in the Peruvian upwelling system, *Biogeosciences,* 14(20), 4601–4617. *https://doi.org/10.5194/bg-14-4601-2017*

de Souza, G.F., Khatiwala, S.P., Hain, M.P., Little, S.H. (2018) On the origin of the marine zinc-silicon correlation, *Earth and Planetary Science Letters,* 492, 15–34. *https://doi.org/10.1016/j.epsl.2018.03.050*

Guinoiseau, D., Abouchami, W., and Galer, S.J.G. (2018a) Effect of CdS precipitation of Cd isotopes: implications for Cd oceanic cycle, *Geophysical Research Abstracts,* 20, 10835. *https://ui.adsabs.harvard.edu/abs/2018EGUGA..2010835G* (Accessed Nov2020)

Baars, O., Abouchami, W., Galer, S.J.G., Boye, M., and Croot, P.L. (2014) Dissolved cadmium in the Southern Ocean: Distribution, speciation, and relation to phosphate, *Association for the Sciences of Limnology and Oceanography.,* 59(2), 385–399. *https://doi.org/10.4319/lo.2014.59.2.0385*

John, S.G., Helgoe, J., and Townsend, E. (2018) Biogeochemical cycling of Zn and Cd and their stable isotopes in Eastern Tropical South Pacific. *Marine Chemistry,* 201, 256–262. *http://.doi.org/10.1016/j.marchem.2017.06.001*

Butler, E.C.V., Butt, J.A., Lindstrom, E.J., Teldesley, P.C., Pickmere, S., Vincent, W.F. (1992) Oceanography of the Subtropical Convergence Zone around southern New Zealand, *New Zealand Journal of Marine and Freshwater Research,* 26(2), 131–154. *https://doi.org/10.1080/00288330.1992.9516509*

Jacquot, J.E. and Moffett, J.W. (2014) Copper distribution and speciation across the International GEOTRACES Section GA03, *Deep–Sea Research II,* 116, 187–207. *https://doi.org/10.1016/j.dsr2.2014.11.013*

Pelizardi, F., Bea, S.A., Carrera, J., Vives, L. (2017) Identifying geochemical processes using End Member Mixing Analysis to decouple chemical components for mixing ratio calculations, *Journal of Hydrology,* 550, 144–156. *https://doi.org/10.1016/j.jhydrol.2017.04.010*

Roshan, S., Devires, T., Wu, J., Chen, G. (2018) The Internal Cycling of Zinc in the Ocean, *Global Biogeochemical Cycle,* 32(12). *https://doi.org/10.1029/2018GB006045*

Quay, P., Cullen, J. Landing, W., and Morton, P. (2015) Processes controlling the distributions of Cd and PO4 in the ocean, *Global Biogeochem. Cycles,* 29, 830 – 841. *https://doi.org/10.1002/2014GB004998*

Weber T, John S, Tagliabue A, DeVries T. (2018) Biological uptake and reversible scavenging of zinc in the global ocean, *Science,* 361(6397), 72 – 76. *https://doi.org/10.1126/science.aap8532*

Roshan, S., and Wu, J. (2015) Cadmium regeneration within the North Atlantic, *Global Biogeochem. Cycles,* 29, 2082–2094. *https://doi.org/10.1002/2015GB005215*

Cutter, G. A., and Bruland, K. W. (2012) Rapid and noncontaminating sampling system for trace elements in global ocean surveys. *Limnol Oceanogr Methods* **10**, 425–436. *https://doi.org/10.4319/lom.2012.10.425*

Biller, D.V., and Bruland, K.W. (2011) Analysis of Mn, Fe, Co, Ni, Cu, Zn, Cd, and Pb in seawater using the Nobias–chelate PA1 resin and magnetic sector inductively coupled plasma mass spectrometry (ICP-MS), *Marine Chemistry,* 130–131, 12–20. *https://doi.org/10.1016/j.marchem.2011.12.001*

Parker, C.E., Brown, M.T., Bruland, K.W. (2016) Scandium in the open ocean: A comparison with other group 3 trivalent metals, *Geographical Research Letters,* 43(6), 2758–2764. *https://doi.org/10.1002/2016GL067827*

GEOTRACES, International Data Assembly Center, Pacific Ocean (2007) *https://www.bodc.ac.uk/geotraces/cruises/section_maps/pacific_ocean/* (accessed Nov 2020)

GEOTRACES, International Data Assembly Center, Atlantic Ocean (2007) *https://www.bodc.ac.uk/geotraces/cruises/section_maps/atlantic_ocean/* (accessed Nov2020)

Noble, A.E., Echegoyen-Sanz, Y., Boyle, E.A., Ohnemus, D.C., Lam, P.J., Kayser, R., *et al.* (2015) Dynamic variability of dissolved Pb and Pb isotope composition from the U.S. North Atlantic GEOTRACES transect, *Deep Sea Research Part II: Topical Studies in Oceanography,* 116, 208–225. *http://doi.org/10.1016/j.dsr2.2014.11.011*

Guinoiseau, D., Galer, S.J.G., and Abouchami, W. (2018b) Effect of cadmium sulphide precipitation on the partitioning of Cd isotopes: Implications for the oceanic Cd cycle, *Earth and Planetary Science Letters,* 498, 300–308. *https://doi.org/10.1016/j.epsl.2018.06.039*

Bourne, H. L., Bishop, J. K. B., Lam, P. J., & Ohnemus, D. C. (2018) Global spatial and temporal variation of Cd:P in euphotic zone particulates. *Global Biogeochemical Cycles,* 32, 1123 – 1141. *https://doi.org/10.1029/2017GB005842*

Karl, D.M. (2002). Nutrient dynamics in the deep blue sea, *Trends in Microbiology,* 10(9), 410–418. *https://doi.org/10.1016/S0966-842X(02)02430-7*

Author

James Moore graduated in Spring 2021 with a double major in biochemistry and botany with a minor in applied mathematics.

The Geotraces Program

The GEOTRACES program aims to better the understanding of Earth's oceans by extensively studying their composition. The chemical analysis of water samples from around the world generates large data products that detail the makeup of water masses throughout the ocean. This paper aims to explore some of the different methods used to investigate coupling between pairs of analytes. Coupling suggests a possible underlying mechanism, and the extent of coupling between two nutrients needs to be quantified to further consider the root of the cause. Using a data product generated by the GEOTRACES GP16 cruise (Eastern Pacific Zonal Transect), the relationship between pairs of macronutrients and trace elements are compared to an oxygen minimum to relate oxygen concentrations to the extent of nutrient coupling. In order to analyze the extent of decoupling, three types of plots were created: 1) depth profiles, 2) differences between the concentrations after applying a scale factor, and 3) direct comparisons of the analytes. Compiling these methods provides a better visualization of transect data, and it is pertinent to research that investigators can identify the location and extent of decoupling events out of massive data sets. Calculating a difference between the coupled analytes after applying a scaling factor provided the most useful information about decoupling.

IdeaFest
Journal

The Student News Digital Archive at Cal Poly Humboldt

Garrett A. Purchio, Cal Poly Humboldt

Abstract

The Cal Poly Humboldt Library has made available two online archives for student newspapers at the university; The Lumberjack and El Leñador. The newspaper archives, which chronicle the history of the university and local communities since 1929, enable global access to stories, photographs, and more as told through the lens of student journalists and contributors. This article chronicles the efforts of the Cal Poly Humboldt Library to create these online archives.

Keywords: Student Publications, Online Archives, Cal Poly Humboldt

Introduction

Digital archives for The Lumberjack and El Leñador, two award-winning student publications, are available online thanks to the Cal Poly Humboldt Library. Over the course of three years, the Cal Poly Humboldt Library worked to make all editions of both publications available online for researchers and alumni. Prior to the creation of these archives, the newspapers were only accessible in bound volumes held within the library's Special Collections & Archives. These collections represent a significant development for research using primary sources by the campus community and researchers in general. The creation of these archives aligns with the library's efforts to provide information literacy instruction on using primary sources for research while also aligning with larger campus curricular goals and needs.

Background

The Cal Poly Humboldt Library supports the curriculum of departments across the three colleges of Cal Poly Humboldt. This support includes providing access to a variety of physical and electronic resources, including newspaper articles, to support the research needs of the students and faculty in these departments. Such support is outlined in the mission statement of the Cal Poly Humboldt Library, which states, "[t]he mission of the Library at Humboldt State University is to provide the information, collections, and services that are necessary to support the instructional programs, research, and outreach services of Humboldt State University," (Cal Poly Humboldt Library, n.d.)

Access to newspapers is provided through online databases that the library subscribes to as well as physical editions that are available for use through the library's print holdings, microfilm collection, bound volumes, and some physical holdings of historic newspapers in Special Collections and Archives. Librarians work with faculty to provide information literacy instruction on finding newspaper articles to students in credit-bearing courses. Students meet with librarians at the library's Special Collections and at the Research Help Desk (RHD) to receive assistance for finding newspaper articles on any number of research topics.

Literature Review

The literature stresses the value of student access to primary sources given the implications for both instruction and scholarship, while also highlighting challenges associated with providing access.

Several authors described the benefit that digitized primary sources will have in information literacy instruction. Murphy surveyed faculty at Maynooth University to assess their perception of the value of digitized primary sources. The survey demonstrated that digitized primary sources had their merits and drawbacks, with many faculty placing value on students being able to access and use original primary sources (Murphy, 2021). Likewise, Terras noted that access to digitized primary sources has the potential to greatly expand research opportunities in the arts, humanities, and social sciences (Terras, 2015) Billeaudeaux and Scott surveyed students in a credit bearing course at the University of Memphis and determined that the ACRL Framework can be used to assess primary source instruction (Billeaudeaux & Scott, 2019).

Despite a greater understanding of the value of primary sources in information literacy instruction by stakeholders, challenges persist with the use of primary sources in instruction. Hensley concluded that additional information literacy is needed to properly support students conducting archival research beyond one-shot sessions (Hensley et al., 2014). Dreyer and Nofziger discussed multiple challenges with access to primary sources, including but not limited to physical, technological, and cultural barriers (Dreyer & Nofziger, 2021). Yaco, noting that discovery and awareness of primary source collections impacts usage, proposed a method for linking curricula with library collections to address faculty unfamiliarity with library holdings that would, "increase utilization of specialized primary source holdings in a broad range of academic disciplines" (Yaco et al., 2016). Furthermore, Martzoukou wrote that with the Covid-19 pandemic forcing the delivery of instruction and services online that there is additional pressure on academic libraries to make primary sources and other materials available to constituents that have come to expect immediate digital access to library resources (Martzoukou, 2021).

A History of The Lumberjack

The Lumberjack Newspaper is the longest-running student-run newspaper of Cal Poly Humboldt. The newspaper published its first edition in October 1929, when it was first known as the HSTC Rooter. HSTC was the acronym of the institution at the time: Humboldt State Teachers College. A year later, the newspaper dropped this name in favor of the Humboldt Lumberjack. This name stuck for a little over two decades before the paper transitioned in 1951 to the title of The Lumberjack, which has been in use ever since. There was a student publication at the institution in the mid 1920's, but it is not included in the online archives.

Through its nine-decade history, The Lumberjack has been a student voice for aspiring journalists that has chronicled the people, places, and events of the university and surrounding communities. The number of editions printed each academic year has varied. Prior to the university's switch from the quarter schedule to the semester schedule in the 1980s, the paper published up to 40 editions regularly between September and June of the academic year. With the switch to the new system in the 1980s, the publication dates have typically run from the end of August until May just before the annual commencement ceremonies and usually include 14 editions per semester. Students can enroll in one of two courses to earn units for working on The Lumberjack. These courses are currently JMC 327: Lumberjack News Workshop or JMC 427 Advanced Lumberjack News. In addition to the print version that appears on newsstands each Wednesday, the staff also publishes content, including multimedia, to the publication's website: https://thelumberjack.org/.

The Lumberjack Newspaper received multiple awards at the Spring 2022 Associated College Press (ACP) statewide college journalism conference. Among the accolades received were: Best News Series, Best Column, Best Feature Photograph, and Best Editorial Cartoon, along with second and third place in additional categories (Marketing & Communications, n.d.). The newspaper has regularly received recognition by different student media organizations including, but not limited to, the California College Media Association

A History of El Leñador

El Leñador is the university's bilingual publication. Established in 2013, this monthly publication features content in Spanish and English. Andrea Juarez is the current faculty adviser for the publication. The mission statement for El Leñador reads, in part, "Our staff cover and provide news to Latinx and other diverse communities on campus and in Humboldt County. We are committed to providing relevant news and expanding the representations and stories told about people of color and other marginalized groups. Our work helps create more social, political and cultural diversity in local media" (*Our Mission*, n.d.).

While the first edition was published in May 2013, the work to make El Leñador a reality began in 2012. According to an article by student journalist Lashay Wesley, then faculty adviser Marcy Burstiner and Rosamel Benavides-Garb, current Associate Dean of the College of Arts, Humanities, & Social Sciences and former Chair of the Cal Poly Humboldt Department of World Languages and Cultures, met with students in JMC and WLC for nearly a year to plan out the first publication (Wesley, 2013). Thanks to a grant from the College of Arts, Humanities & Social Sciences, the newspaper had the funding to print a four-page edition for a full academic year.

The first editions of El Leñador ran as inserts in the Lumberjack Newspaper. A little over a year after its establishment, El Leñador became a standalone publication that continues to publish separately from The Lumberjack. Students can enroll in JMC 160 El Leñador or JMC 360 Advanced El Leñador to earn units for working on the publication. Following the print publication of each edition, a digital file is placed online on the Issuu website for El Leñador: https://issuu.com/ellenador. Additionally, content is also published to the newspaper's website http://www.ellenadornews.com/.In Spring 2022, El Leñador added to its long list of accolades by winning several awards for student journalism in California. El Leñador took home the Best of Show People's Choice: Newspaper Award from the Associated College Press. For awards handed out by the California College Media Association, El Leñador took home awards for Best Illustration, Best Arts & Entertainment Story, and Best Special Issue/Section, along with second and third place in additional categories (Marketing & Communications, n.d.).

An Overview of The Student News Digital Archive Project

In the Spring 2022 semester the Cal Poly Humboldt Library completed the Student News Digital Archive, a project that began in spring 2019 under the title of the Lumberjack Newspaper Digitization Project. The initial goal of the project was to make available through Digital Commons, the Cal Poly Humboldt Library's website for open access education resources, all editions of the Lumberjack Newspaper from its first published edition in October 1929 to the present. The archive would also afford the Cal Poly Humboldt Library and the Department of Journalism & Mass Communication the opportunity to add future published editions of the student newspaper to the online archive.

The Student News Digital Archive began thanks to a successful funding proposal from the GI2025 initiative. The funding was used to hire and train four student assistants to scan microfilm copies of the Lumberjack Newspaper and apply Optical Character Recognition (OCR) to those scans. Interviews were conducted in January 2019, and the students began the project in early February 2019. Over the course of the spring semester, editions of the Lumberjack spanning 1929 to 1962 were digitized using in-house scanners. The content was made text-searchable by applying OCR using the Abbyy Finereader program. Near the end of the semester, the initial batch of editions were uploaded thanks to the help of multiple librarians. At the ideaFest celebration that May, the project was shared with attendees via a research poster along with a live presentation at a donor luncheon in the Center for Teaching & Learning classroom. In the weeks after the conclusion of the semester, another batch of editions was uploaded to the Digital Commons website.

After the initial phase of the project, the faculty adviser to El Leñador met with the College of Arts, Humanities and Social Sciences Librarian to advocate for the creation of a digital archive for the bilingual publication. It was agreed that a separate archive for this publication was needed. With the addition of the second newspaper, the overall project was rebranded from the Lumberjack Newspaper Digitization Project to the Student News Digital Archive. The name change reflected an emphasis on student news publications at Cal Poly Humboldt in general rather than focusing on one publication in particular. Additionally, the rebrand would allow for the creation of additional online archives for student publications such as The Osprey Magazine in the future, of which there has already been an expressed interest.

In Fall 2020, the project restarted thanks to the Dean of the Library, Cyril Oberlander. A new strategy was devised to export the digitization and application of OCR to an outside vendor to complete these critical steps. The library would still require students to upload the files to Digital Commons, but the outsourcing of the scanning and OCR work would enable the library to save costs and complete the project far quicker than it would have using the previous model. In early 2021, the Cal Poly Humboldt Library received the scanned copies of The Lumberjack from the vendor. More recent editions of the publication were acquired directly from the Issuu website (https://issuu.com/hsulumberjack) used to host current editions of the newspaper. Two student assistants from the Library's Access Services Department were recruited to upload the editions to Digital Commons. This process included not only uploading files to Digital Commons, but also data entry of various metadata about each individual edition. This included pieces of information such as the volume and edition number as well as the total number of pages per edition.

In addition to the editions of the Lumberjack, the two student assistants also uploaded editions of El Leñador to the bilingual publication's online archive. As with the Lumberjack, metadata was applied to each edition by the students. Unlike the Lumberjack, however, the editions for El Leñador were not scanned from microfilm, but were instead downloaded directly from the publication's Issuu website (https://issuu.com/ellenador). Since the files were taken directly from what was sent to the printer, there was no need to have OCR applied to the files. Over the course of the next year, these files were uploaded to the online archive with the last edition being uploaded just before noon on February 10, 2022. At the 2022 Authors Celebration, an annual showcase of the published works of Cal Poly Humboldt authors, a formal announcement was made to attendees that the archives are fully online. Additional outreach and marketing of the newspaper archives has included social media posts, a short article on the library's homepage the week following the Authors Celebration, as well as the development and offering of a SkillShop on archival research using historic newspapers.

Newspaper Archives Support Information Literacy Instruction at Cal Poly Humboldt

The Student News Digital Archive enables the Cal Poly Humboldt library to offer a robust research resource to the campus community at a time where primary sources are utilized for research in a number of contexts, and the library is seeing a greater demand for the delivery of information literacy online. Librarians at Cal Poly Humboldt offer information literacy instruction to students in credit bearing courses across the university's three colleges. Instruction is offered both synchronously and asynchronously. Synchronous instruction was, prior to the Covid-19 pandemic, offered primarily through one-shot sessions held in the library, but librarians would also go to classrooms across campus to work with students in credit-bearing courses. With the return to in-person learning in Spring 2022, synchronous instruction has resumed in classrooms across the library and campus, but is also available virtually through the Zoom web application.

Requests for instruction on research using primary sources come from disciplines across the university. These sessions prepare students to complete assignments where they are required to find, utilize, and cite primary source materials. From the 2017-2018 to 2019-2020 academic years, librarians offered an estimated 47 instruction sessions per calendar year on using primary sources. This figure includes sessions requested by faculty that teach credit-bearing courses at the university, community organizations including local schools, as well as 50-minute drop-in sessions offered by librarians as part of the Cal Poly Humboldt Library's SkillShops program. These sessions, a majority of which were offered by three Cal Poly Humboldt Librarians, utilize online and physical primary sources. An emphasis is placed on the materials held with the Cal Poly Humboldt Library Special Collections & Archives, whose primary areas of specialization are Native Peoples, primary industries, and natural resources of Northwestern California as well as the history of Cal Poly Humboldt. In recent semesters, primarily due to the pandemic, instruction sessions on primary sources have been held online using Zoom. Additionally, one Skillshop and multiple course-specific modules have been developed to introduce learners to primary sources using Cal Poly Humboldt's Special Collections & Archives, as well as external online resources.

In addition to meeting an important curricular need, the completion of the online archives comes at a time where the library is seeing a dramatic shift to online instruction. Over the past five years the library at Cal Poly Humboldt, in addition to campus partners, have developed and offered asynchronous learning modules in the Canvas Learning Management System. While some course-specific modules have been developed, the majority of learners that utilize information literacy Canvas modules do so using the library's online SkillShops program. SkillShops began in the 2013-2014 academic year as a series of free workshops for learners from the campus and community–primarily university students–to develop academic, personal, and professional skills.

These asynchronous SkillShops are self-paced modules that students complete on their own timeline. Faculty in credit-bearing courses have assigned SkillShops to students to learn various information literacy skills including, but not limited to, basic and advanced research strategies, how to properly cite sources, and how to evaluate different types of sources, among others. Upon completion, learners are issued badges through the Badgr software program. In Spring 2022, over 1,000 badges were awarded, while in Fall 2021 the program issued 970 badges to learners that completed Canvas SkillShops–a then-record for the online instruction modality that began in the Fall 2017 semester, with Research Basics being the first one offered. As of the end of the Spring 2022 semester, there are 18 asynchronous modules that learners can choose from, which cover topics ranging from citation styles to digital media project and research skills development. In that semester alone, students in at least eight different courses were assigned to complete 24 SkillShops as part of their course assignments.

Going forward, the need for digitized primary sources will only continue to grow. In the Polytechnic Prospectus, the university laid out plans to increase student enrollment while maintaining that about 30% of the total course offerings will be

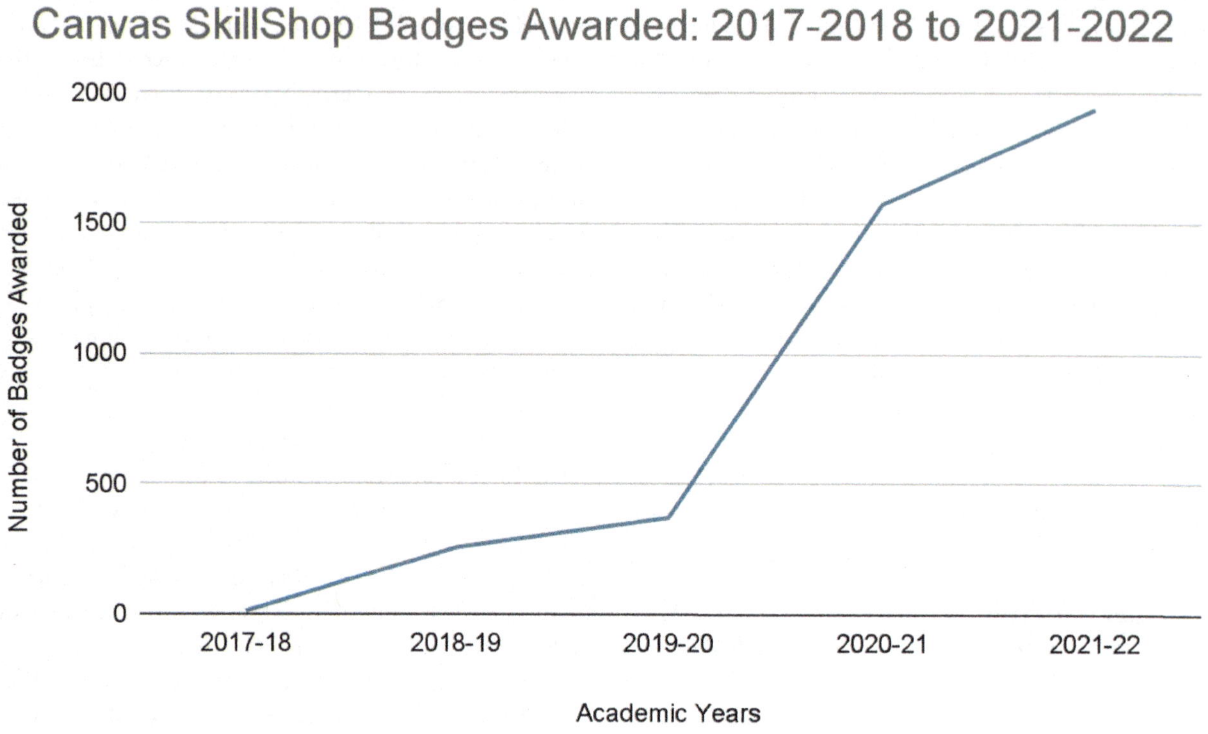
Canvas SkillShop Badges Awarded: 2017-2018 to 2021-2022
2000
1500
1000
500
0
Number of Badges Awarded
2017-18
2018-19
2019-20
2020-21
2021-22
Academic Years

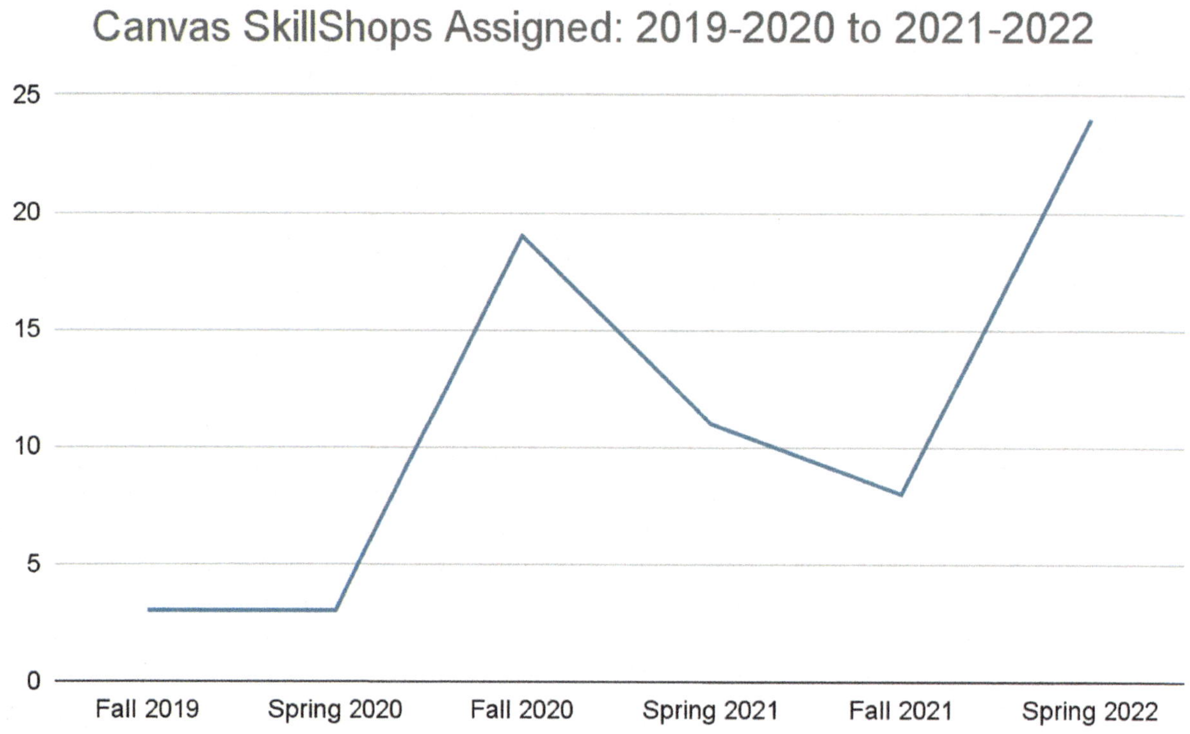
Canvas SkillShops Assigned: 2019-2020 to 2021-2022
25
20
15
10
5
0
Fall 2019
Spring 2020
Fall 2020
Spring 2021
Fall 2021
Spring 2022

online (2021). Such a percentage of online students necessitates immediate access to primary sources that will meet academic, professional, and personal research interests and needs. In addition to the needs of the campus, the communities around the university continue to be avid users of primary sources held by Cal Poly Humboldt. The Special Collections & Archives regularly hosts researchers from the local community, as well as from around the world, to utilize its vast collections of one-of-a-kind primary sources. One local event that sees usage of the primary sources contained within the library's Special Collections and Archives is the annual Humboldt County History Day. This annual spring event enables local students from fourth grade through high school to conduct original research on a selected theme and then present their work on campus ("Humboldt County History Day," n.d.)

Future of the Student News Digital Archive

As was previously mentioned, the plan going forward is to continue to add new editions of both The Lumberjack and El Leñador to their respective online archive collections at the conclusion of each semester. No further scanning of microfilm will be necessary as the digital files that are posted to each publication's Issuu website will be acquired and added to the respective archives in Digital Commons. These born-digital files will not need to have OCR applied to them prior to being placed online. The archives will remain on the Digital Commons platform, barring any need for the location of the holdings to change.

There is interest in creating an online archive for the Osprey Magazine, another student-run publication that has been around since 1975. As with the Lumberjack, the editions for The Osprey will need to be scanned from microfilm and have OCR applied to the pages before they are placed onto Digital Commons. There is currently no timetable when this student publication would be scanned from microfilm and have its own archive created on Digital Commons.

Conclusion

The Student News Digital Archive opens up opportunities for accessing one-of-a-kind primary sources while also enabling the library to meet the need for instruction on using primary sources in an increasingly online environment. Statistics from Digital Commons will be able to indicate how often the archives are utilized and where users are accessing and downloading editions from. Alumni, current and future researchers, as well as anyone with an interest in the history of Cal Poly Humboldt through the lens of the students who have studied here for over a century, can access the archives from anywhere in the world thanks to the efforts of library student assistants and employees.

Thanks to Everyone Involved

The Student News Digital Archive is a collective effort by the following individuals:

Cal Poly Humboldt Library Student Assistants

Kayla Garcia, Lisa Salcido, Veronica Koomson Maiden, Danielle Kirkland-Shatraw, Reanne Meighan, and Adam Mellot.

Cal Poly Humboldt Library Employees

Kyle Morgan, Bri Hagen, Janee Mason, Keelye Wilson, Jessica Welch, Cyril Oberlander, Andy Pierce, and Carly Marino

Department of Journalism and Mass Communication

Vicky Sama, Deidre Pike, Andrea Juarez, Marcy Burstiner, and Grace Caswell

To everyone involved with this project, thank you for all you have done to make over 90 years of university history accessible to researchers and alumni everywhere.

Bibliography

Billeaudeaux, B., & Scott, R. E. (2019). Leveraging Existing Frameworks to Support Undergraduate Primary Source Research. *Reference & User Services Quarterly*, *58*(4), 246–255.

Cal Poly Humboldt Library. (n.d.). *Library Mission and Vision.* Retrieved June 7, 2022, from https://library.humboldt.edu/about/mission.html

Dreyer, R., & Nofziger, C. (2021). Reducing Barriers to Access in Archival and Special Collections Public Services. *Pennsylvania Libraries*, *9*(1), 37–48. http://dx.doi.org/10.5195/palrap.2021.237

Hensley, M. K., Murphy, B. P., & Swain, E. D. (2014). Analyzing Archival Intelligence: A collaboration between library instruction and archives. *Communications in Information Literacy*, *8*(1), 96–114. http://dx.doi.org/10.15760/comminfolit.2014.8.1.155

Humboldt County History Day. (n.d.). *Humboldt County Office of Education.* Retrieved May 8, 2022, from https://hcoe.org/history-day/

Humboldt State University. (2021). *Polytechnic Prospectus* (p. 144) [Prospectus]. Humboldt State University. https://www.humboldt.edu/sites/default/files/hsupolytechnic-selfstudy.pdf

Marketing & Communications. (n.d.). *Newspapers, Magazine Recognized for Powerful Journalism.* Humboldt Now. Retrieved April 20, 2022, from https://now.humboldt.edu/news/newspapers-magazine-recognized-powerful-journalism

Martzoukou, K. (2021). Academic libraries in COVID-19: A renewed mission for digital literacy. *Library Management, 42*(4/5), 266–276. https://doi.org/10.1108/LM-09-2020-0131

Murphy, H. (2021). Understanding the Value of Digital Archival Collections to Faculty at Maynooth University Library. *New Review of Academic Librarianship, 27*(4), 423–439. https://doi.org/10.1080/13614533.2021.1976233

Our Mission. (n.d.). El Leñador. Retrieved February 18, 2022, from https://www.ellenadornews.com/about-us/our-mission/

Terras, M. (2015). Opening Access to collections: The making and using of open digitised cultural content. Online Information Review, 39(5), 733–752. http://dx.doi.org/10.1108/OIR-06-2015-0193

Wesley, L. (2013, May 8). El Leñador: The Making of a Bilingual Publication. El Leñador.

Yaco, S., Brown, C., & Konrad, L. (2016). Linking Special Collections to Classrooms: A Curriculum-to-Collection Crosswalk. *American Archivist, 79*(2), 417–437. https://doi.org/10.17723/0360-9081-79.2.417

IdeaFest
Journal

The Influence of Hiking Trails on Salamanders

Emily M. MacDonald

Abstract

Over half of the World's salamanders are facing extinction. It is important to understand the impact Humans have on them in all parts of their ranges so we can better understand what they need and how we can prevent further extinctions. I conducted this study to try to determine if the human activity on the official hiking trails in the Arcata Community Forest in Arcata California; as designated by the official map of the Forest; is negatively impacting the salamanders that live in the Forest. I thought there would be an increase in the number of species and the overall number of salamanders as the plots got further away from the hiking trails. I conducted area constrained searches to examine the individual abundance and species diversity of salamanders in the Arcata Community Forest to their distance from the hiking trails. I also examined the correlation between the salamanders weight and snout-vent length to the distance from the trail the salamanders were found at. The salamanders were located and measured using a series of area constrained searches inside 1x1 meter plots, with a distance of at least 20 meters between them. The plots covered an area from 0 to 11 meters away from the trail. The salamanders were bagged, identified to species, then weighed and measured from their snout to their vent. The data was analyzed using Spearman's Rank Correlation analyses and used t-values to determine the significance of the correlations. The p-values for the Spearman's Rank Correlations for the species number to the distance from the trail and the number of individual salamanders to the distance from the trail both came back as having significant correlations with higher numbers of species and individuals closer to the trails. This could mean the trails are acting as ecological traps for the salamanders, so the trails and their associated human activities may still be harming the salamanders, despite the salamanders seemingly being drawn to the trails instead of staying away from them like I thought they would. Follow up studies are needed to determine if this is a case of an ecological trap.

Keywords: Salamander, Diversity, Hiking trail, Humboldt County, California, Community forest, Area constrained search

With over 66% of the Earth's salamander species being listed at the level of near threatened or worse, we may soon lose many, if not all of these species to extinction without intervention (IUCN 2022). The United States Fish and Wildlife Service states that if current trends continue amphibians will disappear from half of their habitat in the next 30 years (USFWS 2015).

Past studies have shown that amphibian diversity is lower in urban areas but that can vary based on how much human activity there is and what kind of activity it is (Janiawati et al. 2015). With the global decline of amphibians (IUCN, and Conservation International 2008) it is important to get baseline data so that there are numbers to compare future studies to, that way any changes in species composition or abundance can be noticed and addressed sooner.

Hiking trails are very popular for outdoor recreation, and they often occur near or alongside urban areas. The Arcata

Community Forest is home to many hiking trails that are greatly enjoyed by both locals and tourists. Many people benefit from them, but is the same true for the salamanders who call the forest home, or are they affected by the hiking trails in the same way urban activity affects them, with lower diversity in areas of Human activity. Salamanders are occasionally found dead on these trails after being stepped on or run over by people on bikes as they try to cross them to move between patches of forest. If the human activity on the hiking trails is negatively impacting the salamanders, there will be an increase in the levels of species diversity and the overall number of salamanders as the plots get further away from the hiking trails. If the distance from the hiking trails impacts the salamanders than there will be a significant correlation between the distance to the hiking trails and the size of the salamander.

Study Area

This study was conducted in the Arcata Community Forest, Humboldt County, CA, USA. There are many species of salamanders in Humboldt County. Of these species, 10 occur in and around the Arcata Community Forest (McGinnis and Stebbins 2018, AND Reilly and Wake 2019). I found three of them during the study. These species were the Northwestern Salamander (*Ambystoma gracile*) (Figure 6), Common Ensatina (*Ensatina eschscholtzii*) (Figure 4), and the California Slender Salamander (*Batrachoseps attenuates*) (Figure 5) (McGinnis and Stebbins 2018).

The salamanders in this area are subject to predation by vertebrates such as Garter Snakes (*Thamnophis elegans and Thamnophis sirtalis*) (California Herps 2020), birds (Fitzpatrick et al. 2009), and each other (McGinnis and Stebbins 2018).

The trees in the forest are primarily coniferous trees such as Coast Redwoods (*Sequoia sempervirens*) (Visit Redwoods 2021). The understory consists primarily of Redwood Sorrel (*Viola sempervirens*), Redwood Violets (*Oxalis oregana*), Pacific Trillium (*Trillium ovatum*), and Western Sword Ferns (*Polystichum munitum*). The forest is used for many recreational activities. There is a disc golf course and several mountain biking courses. The forest is also popular for walking dogs, riding horses, and hiking. There is a playground in Redwood Park, which is adjacent to the community forest. There is also a lot of questionable anthropogenic waste, and a lot of dogs are allowed to run off leash. There are also structures such as Jolly Giant Reservoir that are no longer in use. There are also water storage tanks and large power lines that cut through part of the Arcata Community Forest (Trail Maps: Arcata, CA. 2021). There are instances where salamanders are found dead on the trails after being run over or stepped on (iNaturalist AND personal observations.).

Methods

The study took place from March 5th, 2022, through April 11th, 2022. During the course of this study, I compared the individual abundance and species diversity of salamanders in the Arcata Community Forest to their distance from the official hiking trails. I also examined the correlation between the salamanders' measurements of weight and snout-vent length with the distance from the trail the salamanders were found at.

The study was conducted by setting up plots in which area constrained searches (PARC 2013) took place. To assign the distance from the trail each plot would be at randomly, I used the RANDBETWEEN function in Microsoft Excel with a range of 0 to 10. The distances were measured from the edge of the trail to the center point of the plot's closest edge. The plots alternated which side of the trail they were on to increase the potential number of microclimates that could be sampled, the starting side was also chosen randomly with a coin flip. The plots started at distances of 0 to 10 meters away from the trail. The plots were at least 20 meters apart at the closest point between them. Once I determined where the plot was going to be, I measured out a one-meter length with the center point of it on the line from the trail. Then I placed a flag on each end and used them as points to measure from when placing the last two flags. Then I double checked the measurements and angles to make sure the plot was the correct size before starting the search. I also placed a thermometer in the center of the plot to take the temperature. Then I put on some nitrile gloves and began the search. The gloves protected the salamanders from any oils or bacteria present on human skin and changing them between salamanders helped to mitigate the risk of spreading disease between individuals. The searches themselves consisted of looking through leaf litter (primarily redwood duff with some deciduous leaves), and lifting logs, rocks, fern fronds, and pieces of bark. Salamanders have very sensitive skin, so they seek shelter in cool damp places such as leaf litter, under or within logs and under pieces of bark (Pough 2007).

Any salamanders that I found were individually placed in separate unsealed bags and kept upright in a small bucket. I then returned the cover object to its original position, this way when I was done working with each salamander, I could return them to their cover object more quickly and the salamanders could find their way back under their cover object safely. If I were to place the cover object back down on top of them, they could be injured or killed. Once they were in the bags, I took their photo and identified them to species. Next, I weighed them by clipping the bag onto a small portable hanging scale. Then I used the bags to gently restrain them so I could get the snout to vent length measurement without them moving as much. If at any point the salamander I was working on were to have shown signs of stress, including secreting toxins, doing the unken reflex, biting, or dropping its tail; I would have placed it back down next to its cover object immediately regardless of what point I was at in the data collection process.

If a cover object such as a log or rock was partially inside a plot, I measured it in the longest direction, and if 50% or more of its length was in the plot it was searched, and any salamanders found under it were included in the count for that plot. Any salamanders that were found under the part of the log that was out of the plot were still counted, and were moved to the correct distance category in the data sheet. If less than half of the log was in the plot but there was a salamander under the potion of the log that was in the plot it was still included in that plot's data set. Any salamanders that were found in the plot that were not under a cover object or that walked into the plot while I was searching it, were still included as a part of that plot's data.

I used Spearman's Rank Correlations to look at the distance from the trail to the number of salamanders and the number of species found at each distance. I also used it to look at the distance from the trails and the snout to vent length and weight of each species. I used the degrees of freedom, and both one and two-tailed t-tests to get the numbers to plug into the equations to be able to calculate the P-values.

Results

In total I was able to search 25 plots. I found one Northwestern Salamander (*Ambystoma gracile*), eleven Common Ensatinas (*Ensatina eschscholtzii*), and nine California Slender Salamanders (*Batrachoseps attenuatus*), for a total of 21 salamanders and three species observed out of the 10 species that have been reported as occurring in or near the Arcata Community Forest. The Spearman's Rank Correlation between the number of species and the distance away from the trail has a negative correlation of -0.615 and it has a p-value of 0.029. The Spearman's Rank Correlation between the number of individual salamanders and the distance away from the trail has a negative correlation of -0.6, and it has a p-value of 0.033. There is a significant correlation between the distance away from the hiking trails in the Arcata Community Forest and both the number of salamander species and the number of individual salamanders. However, the correlations are both negative. This means that there are more species and more individuals closer to the hiking trails instead of further away from them, like I predicted in my hypothesis (Figure 1). The results for whether or not there was a significant correlation between the distance from the trail and the weight and snout to vent lengths for the Northwestern Salamander was not tested because only one individual was found. The results for whether or not there was a significant correlation between the distance from the trail and the weight and snout to vent length for the Common Ensatinas came back as not significant. For the correlation between the distance from the trail and the salamanders weight the Spearman's Rank Correlation 0.136 and the p-value is 0.707. For the correlation between the distance from the trail and the salamanders snout to vent length the Spearman's Rank Correlation is -0.109 and the p-value is 0.764 (Figure 2). However, both data sets came back as nearly significant for the California Slender Salamanders with positive correlations of 0.512 for the distance from the trail and their weight, and a positive correlation of 0.133 for the distance from the trail to their snout to vent length. Both measurements have near significant p-values of 0.05 (Figure 3).

Discussion

My first hypothesis; if the human activity on the hiking trails is negatively impacting the salamanders, there will be an increase in the levels of species diversity and the overall number of salamanders as the plots get further away from the hiking trails; was correct in that there was a significant correlation, however it was wrong in the direction the correlation went. The number of individual salamanders and the number of species were both higher closer to the trails. There are several possible explanations for these results.

The hiking trails or the habitat next to them may be acting as ecological traps. An ecological trap is a low-quality habitat or a habitat that negatively impacts the survival of the species in question but is preferred over high-quality habitat that would improve survival by the species in question (Battin 2004). The hiking trails seem at first glance that they would be the part of the forest that is the most used and therefore the most damaged, however the way people use the Arcata Community Forest is a lot more complex than I realized before going into the field to gather data. There appears to be more variables at play than simply the distance from the hiking trails. There were some areas off-trail that had large piles of garbage, indicating that people had spent some time in these off-trail areas and had not been treating the habitat with respect. The off trail human activity may be causing more environmental damage and having a higher negative impact on the salamanders, perhaps those that stick to the trail are more conscious of their impact on the surrounding area and cause less damage to it. The quality of the habitat next to the trails could be of better quality despite the risk of being runover by bicyclists or stepped on by hikers. More research needs to be done to determine if the habitat near the trails is an ecological trap or if the salamanders are using it because it is better than the habitat off-trail.

Only finding three of the ten species that have been reported as living in the Arcata Community Forest could be due to several factors. If the habitats by the trails are ecological traps more sensitive species may have died out or become rarer in these areas. I also could have missed them due to the randomness of where the plots were placed, or they were in areas I did not search. There were some areas that were simply inaccessible or that posed a safety risk. There were many steep slopes, rocky areas, and dense thickets that I could not access with the equipment that I had. This means there were some areas in which I could not search for salamanders. Some of the species may be less common or may naturally occur at lower densities then other species of salamander in the Arcata Community Forest. There could have been less salamanders because of lower moisture levels due to the drought.

The state of California has been experiencing less than typical levels of rain and has entered a drought. Some people in Humboldt County have even reported that the wells that supply their homes with water have gone dry (California Drought Update March 21, 2022). The drought may be why there were so few individual salamanders, and so few species of them. Salamanders need some degree of moisture in their habitat, they can die if they get too dry as they have semi permeable skin. Many species use burrows made by themselves or by other animals such as small mammals and go deep underground to escape the heat and dryness (The National Wildlife Federation. n.d). This means they might not have been at the surface for me to find. I do not think breeding season is a big factor in why I did not find some species. It is breeding season for the Common Ensatinas and Northwestern Salamanders and I still found both species during the study (McGinnis and Stebbins 2018). My second hypothesis that "If the distance from the hiking trails impacts the salamanders than there will be a significant correlation between the distance to the hiking trails and the size of the salamander." does seem to be correct for the California Slender Salamanders. The positive correlations and near significance do seem to indicate that the salamanders that are longer and weigh more are found further from the hiking trails and the smaller salamanders are found closer to the hiking trails (Figure 3). If I found more individuals, I do believe the results would have been significant. These results could indicate that younger individuals are using the habitat that is closer to the trail, perhaps this is the result of intraspecific resource competition. The Common Ensatinas data did not show these results, however I think they may still have a similar correlation. I did not find many older adults and I found no new hatchlings. I found mostly hatching year juveniles and mid-sized juveniles, and I only found and a few young adults. I think the ratio of hatching year juveniles and mid-sized juveniles to older adults might have skewed the data. Some juveniles weighed a lot and were short in length and other juveniles that were longer weighed the same (Figure 2). The Common Ensatinas were found at a larger range of distances from 0 to 7 meters from the hiking trails and the California Slender Salamanders were found from 1 to 4 meters from the hiking trails (Figures 2 and 3). If I were to do this study again using the same methods, I would use a larger premade quadrat to use as the boundaries of the plot to streamline the setup process. If I were to change the methods, I would look under as many cover objects as I was able to and once I found a salamander I would take its measurements and measure its distance to the trail. I would also like to do a longer study at a different time of year, and I would like to scale up the project and have a team of people helping me so more ground could be covered. If it was done at a different time of year after the spring flowers have died back, there would be a much lower risk of harming any of the plants.

Management Implications

The results of this study indicate that something may be drawing the salamanders towards the hiking trails. Follow up studies are needed to determine what variable(s) is(are) drawing the salamanders towards the trails, and why they are drawn to said variable(s). Follow up studies are also needed to determine if the areas next to the trails are acting as ecological traps. Longer term studies could also be important for gathering this data on the other seven species of salamanders that have previously been found in and near the Arcata Community Forest but were not found during this study.

Acknowledgements

This study was supported by California State Polytechnic University Humboldt's Wildlife Department.

Literature Cited

Battin, J. (2004). When good animals love bad habitats: Ecological traps and the conservation of animal populations. *Conservation Biology* 18:1482–1491.

California Drought (2022). California Drought Action. Retrieved May 5, 2022, from https://drought.ca.gov/media/2022/03/CA-Drought-Update-3-21-22.pdf

California Herps. (2020). California red-sided Gartersnake - *Thamnophis sirtalis infernalis*. http://www.californiaherps.com/snakes/pages/t.s.infernalis.html

Fitzpatrick, B. M., K. Shook, and R. Izally. (2009). Frequency-dependent selection by wild birds promotes polymorphism in model salamanders. *BMC Ecology* 9.

Janiawati, I. A. A., M. D. Kusrini, and A. Mardiastuti. (2015). Structure and composition of reptile communities in human modified landscape in Gianyar Regency, Bali. *Journal of Indonesian Natural History* 3:85–91.

IUCN, and Conservation International. (2008). Threatened amphibians of the world. IUCN, Conservation International, and Lynx Edicions.

IUCN. (2022). IUCN Red List. Salamanders Stats. https://www.iucnredlist.org/search?query=salamanders&searchType=species&gt

LeGros, D. L., B. Steinberg, and D. Lesbarrères. (2017). Middle of the road: Enhanced Habitat for Salamanders on unused logging roads. *Wildlife Research* 44:1–8.

McGinnis, S. M., and R. C. Stebbins. (2018). *Peterson Field Guide To Western Reptiles And Amphibians*. Fourth edition. Houghton Mifflin Harcourt, Boston, Massachusetts AND New York, New York.

National Oceanic and Atmospheric Administration [NOAA]. (2021). National Weather Service Forecast Office Eureka, CA. https://w2.weather.gov/climate/index.php?wfo=eka

National Oceanic and Atmospheric Administration [NOAA]. (2022). National Weather Service Forecast Office Eureka, CA. Retrieved May 5, 2022, from https://www.weather.gov/wrh/climate?wfo=eka>

iNaturalist. n.d. Observations. Retrieved February 18, 2022, from https://www.inaturalist.org/observations?nelat=40.87791019999999&nelng=-124.04202795&place_id=any&swlat=40.865643&swlng=-124.07563175&taxon_id=26718&view=species.

Pacific Southwest Research Station. (2021). USDA Forest Service. https://www.fs.fed.us/psw/locations/arcata/

Partners in Amphibian and Reptile Conservation, Birmingham (PARC). (2013). Inventory and monitoring: Recommended techniques for reptiles and amphibians: With application to the United States and Canada. Technical Publication IM-1.

Pough, F. H. 2007. Amphibian Biology and husbandry. *ILAR Journal* 48:203–213.

Reilly, S. B., and D. B. Wake. (2019). Taxonomic revision of black salamanders of the *Aneides flavipunctatus* complex (Caudata: Plethodontidae). *PeerJ* 7:e7370.

The National Wildlife Federation. n.d. Tiger salamander. Wildlife Guide. Retrieved May 5, 2022, from https://www.nwf.org/Educational-Resources/Wildlife-Guide/Amphibians/Tiger-Salamander#:~:text=Unlike%20other%20salamanders%2C%20tiger%20salamanders,more%20than%20two%20feet%20deep.

Trail Maps: Arcata, CA.)2021). Trail Maps | Arcata, CA. City of Arcata. Retrieved May 5, 2022, from https://www.cityofarcata.org/197/Trail-Maps

United States Fish and Wildlife Service [USFWS]. (2015). Conserving amphibians: What the amphibians are telling us and why we should listen. https://www.fws.gov/endangered/news/amphibians.html

Visit Redwoods. (2021). Arcata Community Forest. Travel Info for the Redwood Forests of California, Eureka and Humboldt County. https://www.visitredwoods.com/listing/arcata-community-forest/489/

Figures

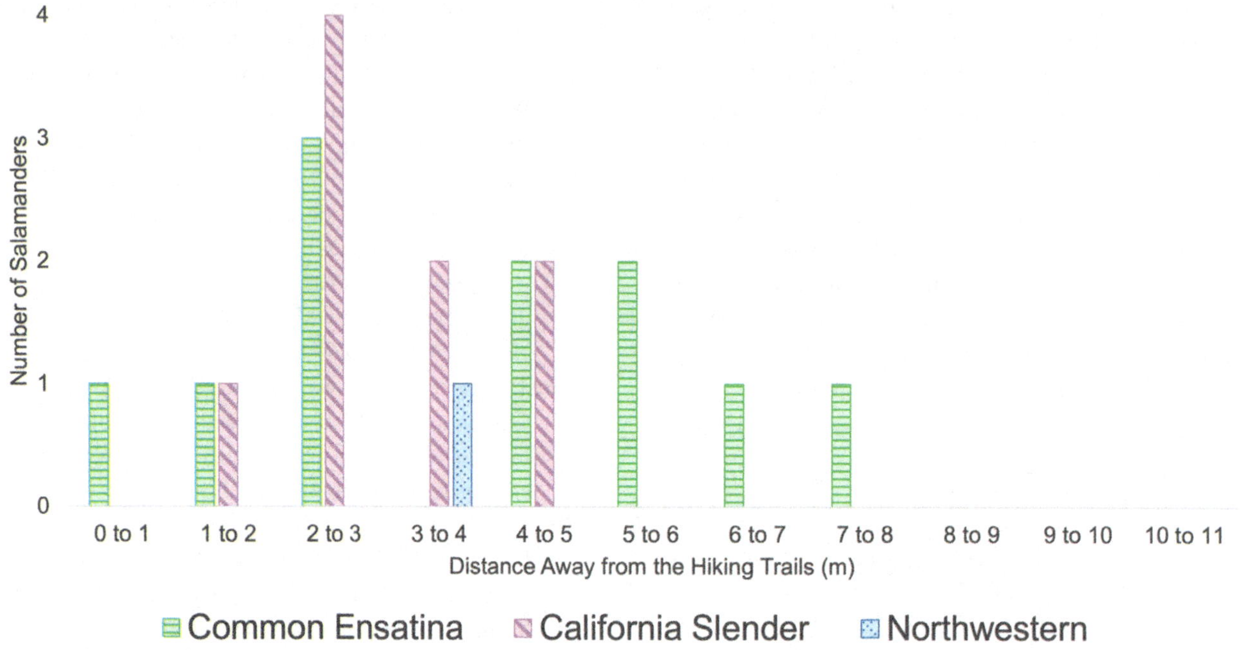

Figure 1. The number of salamanders of each species found at each distance from the hiking trails in the Arcata Community Forest, Arcata California, March-April 2022.

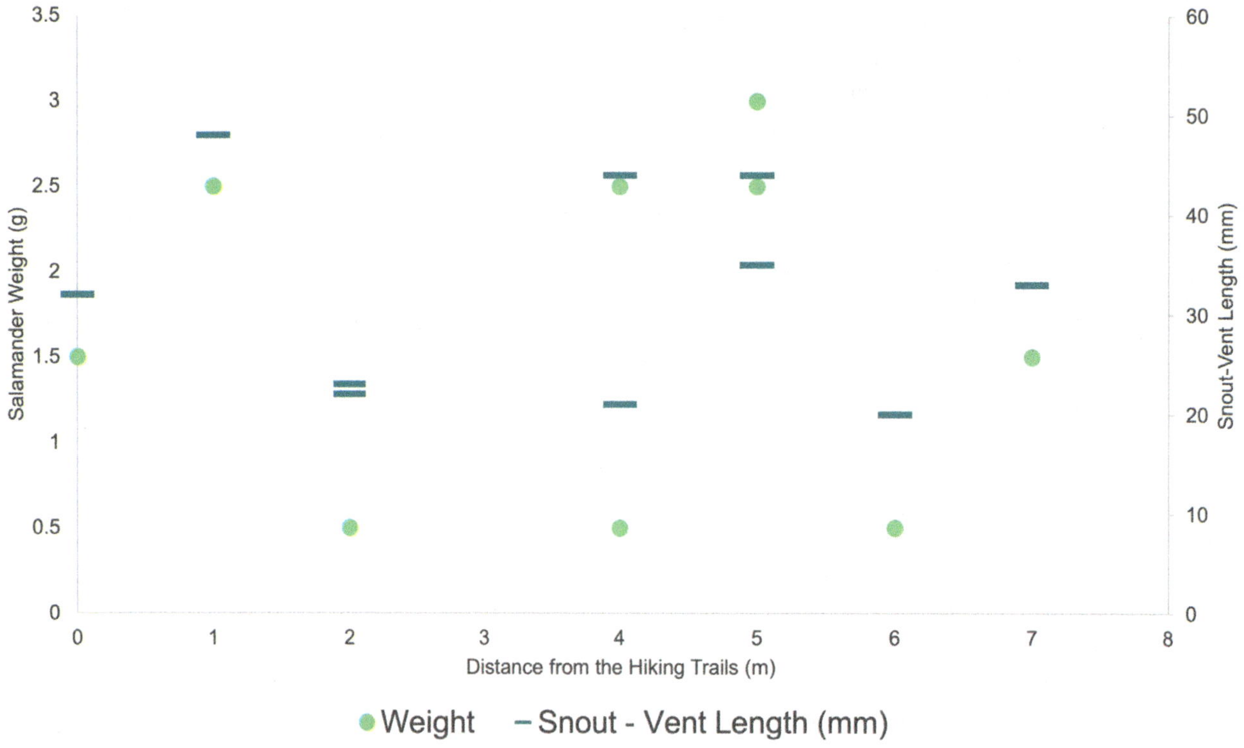

Figure 2. The weight and snout-vent length to the distance from the hiking trails for the Common Ensatinas in the Arcata Community Forest, Arcata California, March-April 2022.

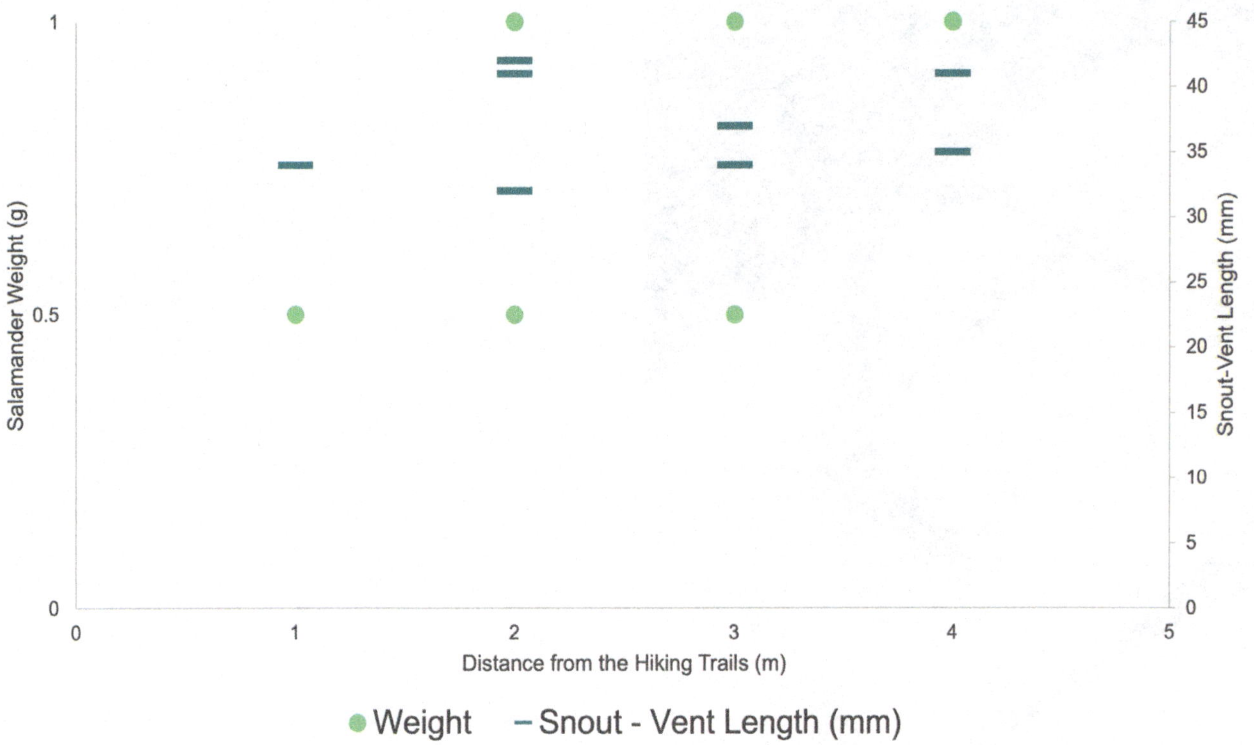

Figure 3. The weight and snout-vent length to the distance from the hiking trails for the California Slender Salamanders in the Arcata Community Forest, Arcata California, March-April 2022.

Figure 4. One of the Common Ensatinas that I found in the Arcata Community Forest, Arcata California, March-April 2022.

Figure 5. One of the California Slander Salamanders that I found in the Arcata Community Forest, Arcata California, March-April 2022.

Figure 6. The Northwestern Salamander that I found in the Arcata Community Forest, Arcata California, March-April 2022.

IdeaFest
Journal

Novel Microbiological Medium Developed For The Isolation Of Bacteria Associated With Estuarine Sea Anemones

Parker K. Lund & Catalina Cuellar-Gempeler

Abstract

Out of the nearly one trillion species of microbiota estimated to inhabit Earth only ten thousand have been cultured in the laboratory. Culturing continues to play a vital role in determining the physiology and ecologic function of individual bacteria in microbial communities and for microbes associated with host organisms one of the major challenges is developing microbiological media that mimics the bacteria's natural environment enough to promote growth. Here, we target bacteria associated with the estuary anemones Diadumene lineata and Metridium senile by developing a novel medium that uses anemone tissue as the sole source of nutrients. We further measured their growth from 30-40°C, identifying candidates for beneficial host-microbe interactions in warm environments. The anemone-based medium overwhelmingly selected for the genera Pseudoalteromonas and Peribacillis, regardless of anemone host species. Peribacillus spp. were particularly thermal tolerant, growing similarly from 30-40°C, while Pseudoalteromonas spp. grew well from 30-35°C. The remaining genera preferred 30°C, however one of the Litoreibacter sp. produced a diffuse putative pyomelanin that may protect cells against thermal stress. Further research should explore the production of secondary metabolites in these isolates, including antimicrobial/antifungal activity that may prevent environmental microbiota from colonizing the host under stressful conditions. This work supports culturing as a critical component of unraveling the individual functions of these bacteria in anemone-associated microbial communities.

Keywords: Microbiological medium, Microbiology, Marine, Estuary, Anemone, Sea anemone, Bacteria, Psuedoalteromonas, Peribacillus, Litoreibacter, Estuarine, Culturing, Isolates

Introduction

Out of the nearly one trillion species of microbiota estimated to inhabit Earth, only ten thousand have been cultured in the laboratory (Locey & Lennon, 2016). Recent diversity estimates are driven by the advance of molecular methods for differentiating taxa (Locey & Lennon, 2016), however culturing continues to play a vital role in determining the physiology and ecologic function of individual bacteria (Joint et al., 2010) including single cell encapsulation in gel micro-droplets. This is important not just for understanding large-scale microbial contributions to ecosystems, like biogeochemical cycling (Hayatsu et al., 2008; Raina et al., 2009; Tortell et al., 1999) but also for applied research like combating pathogens (Hu et al., 2016; Rosado et al., 2019; Santoro et al., 2021)pharmaceutical exploration (Debbab et al., 2010; Joint et al., 2010), and bioremediation (Muriel-Millán et al., 2021). Through culturing, we can also identify the beneficial functions of bacteria in host-associated microbial communities, which produce secondary metabolites (Chen et al., 2019) that can enhance host resilience towards stressors like disease (Lawley et al., 2012; Rosado et al., 2019) and temperature (Cunning & Baker, 2020; Cziesielski et al., 2018; Rosado et al., 2019) For example, targeted probiotic therapies from cultured bacteria have been employed successfully to enhance coral health by mitigating the effects of temperature stress on bleaching and combating pathogens (Rosado et al., 2019). Generating microbe-based therapies first requires identifying the beneficial functions that individual microbiota contribute to the host-microbe system, which can only be achieved through culturing.

Though we can investigate the functional capacity of whole communities through direct measurement, like chlorophyll production, or measuring gene expression via transcriptomics, culturing allows for community manipulation to determine the roles of individual bacteria. Many studies indicate that positive host-microbe interactions are occurring, but it is unclear which bacteria are responsible. For example, in the diatom *Cylindrotheca fusiformis*, the inhibition of growth induced by *Marinobacter* is rapidly reversed by introducing a mixed microbial consortia from the diatom's native environment (Stock et al., 2019). Likewise, in the Hawaiian bobtail squid *Euprymna scolopes*, the accessory nidamental gland becomes stunted or fails to develop entirely without the presence of sediment microbiota from its origin (McAnulty et al., 2023). In contrast, a high-throughput culturing assay using combinations of different bacteria in conjunction with five different carbon sources identified strains that facilitated the growth of the plant symbiont *Herbaspirillum frisingense*. Notably, while one strain of *Burkholderia sp.* was a consistent facilitator regardless of community composition, a *Bacillius sp.* and *Rahnella sp.* only facilitated symbiont growth together in co-culture (Kehe et al., 2019). Because interactions between different microbial taxa can affect their function within the community, these relationships can be extremely complex to disentangle.

One of the major challenges is developing microbiological media that mimics the bacteria's natural environment enough to promote growth, particularly for species that associate closely with a host. Here, "host-associated" refers to bacteria living on or within a another organism, however the strength of these associations vary and the more transient microbes are typically acquired from the surrounding environment (Parfrey et al., 2018). One way of targeting some host-associated taxa is using host tissue as the sole source of nutrients in the medium. For example, producing a medium from yellow onion can be used to culture onion-associated pathogens (Zaid et al., 2012). Likewise, another medium targeted putative fish pathogens and beneficial microbiota by using peptones purified from fish viscera (Vázquez et al., 2004) Kelp-associated bacteria can also be cultured using a medium made out of alginate, the dominate polysaccharide produced by many kelp species such as those in the genus *Macrocystis* (Lin et al., 2018).

In this study, bacteria were isolated from anemones and subjected to a thermal stress assay, identifying heat tolerant individuals that may benefit the host under temperature stress. For targeting anemone-associated bacteria, we used autoclaved anemone tissue as the sole source of nutrients in an agar-seawater medium and attempted to culture bacteria from two species of sea anemone: *Metridium senile* and *Diadumene lineata*. Samples were plated on both anemone medium and marine agar to determine if bacteria could be isolated with anemone nutrients alone, and sequencing the isolates allowed for a comparison between the two media. To identify individual candidates that could contribute to host resilience in warm environments, isolated bacteria were subjected to a heat stress assay measuring their growth over time across a temperature gradient. Although the anemones are closely related and inhabit similar estuary habitats, *D. lineata* displays a greater heat tolerance than *M. senile* (Glon et al., 2019; Shick, 1976). If microbiota are mitigating heat stress in their host, we expect isolates from *D. lineata* to display greater tolerance at high temperatures than isolates from *M. senile*.

Methods

Summary

A microbiological medium was developed using whole tissue from anemones and artificial seawater. Bacteria from *Diadumene lineata* and *Metridium senile* were cultured on both the anemone-based medium and marine agar. Marine agar is a commercially-prepared medium commonly used for culturing marine bacteria. Isolates were selected from both media based on colony morphology, attempting to select as many unique colonies as possible (based on size, shape, color, apparent texture, and opacity). Each isolated colony was grown in its respective broth medium and cryopreserved at -80C using glycerol. For each isolate, the 16S rRNA gene was amplified using single-colony PCR so that they could be identified to genus via Sanger sequencing (in conjunction with the BLAST algorithm). Finally, a subset of isolates were selected for a thermal stress assay to measure their growth at 30°C, 35°C, and 40°C, identifying bacteria that were particularly thermal tolerant. *D. lineata* can survive short exposures at 35°C and 40°C is well beyond the upper temperature limits of host survival, however we also aimed to define the upper limits for these isolates.

Collection of Sea Anemones

D. lineata were identified morphologically by their small size (less than three centimeters in height) and grey-green column adorned with orange/white vertical stripes (Uchida, 1932). Although there are four primary morphotypes (Ryan & Kubota, 2016), they still appear distinct from other anemone species in this region. *M. senile* are larger and average five cm in height, with a smooth column that is commonly white, but can also appear grey, orange, or brown in color (Fautin et al., 1987), however they were predominately white/grey in this collection. They are additionally identified by the collar-like parapet at the top of the column, followed by extensions of feathery tentacles with a frilled margin (Fautin & Hand, 2007). *D. lineata* were collected from the dock at the Arcata Marsh in Arcata, CA (40.856253, -124.098625), and *M. senile* were collected from the docks at the marinas in Eureka, CA (40.803866, -124.176253) and at Woodley Island, CA (40.807440, -124.164895). Small anemones were collected in 50 mL tubes using a spatula and jeweler's forceps (sterilized in 70% ethanol) to carefully pry them from the sides of the docks. Anemones were rinsed to remove as much debris as possible, particularly from the pedal disc, then stored sideways in 15 mL tubes at 4°C in 35 ppt seawater (similar to their areas of origin). They were fed *Artemia* nauplii sporadically and were allowed to feed for a minimum of four hours, giving them time to egest waste before changing the water. This helped further minimize fouling during storage in the refrigerator.

Development of Anemone Medium

Large *M. senile* were collected using a palette knife to gently dislodge them from the bottom of fouled plastic buoys. The anemones were rinsed of excess debris and mucus, before retiring under high pressure at 121°C to breakdown their tissue. The tissue was further homogenized using a coffee grinder and frozen in 50 mL aliquots at -20°C. An initial trial was run testing anemone:water ratios of 1:1, 1:2, 1:3, and 1:4, which all exhibited robust bacterial growth. A ratio of one part anemone tissue for every three parts 35 ppt artificial seawater (by volume) was ultimately selected because the bacterial colonies were moderately sized after 24 hours of incubation. The anemone medium

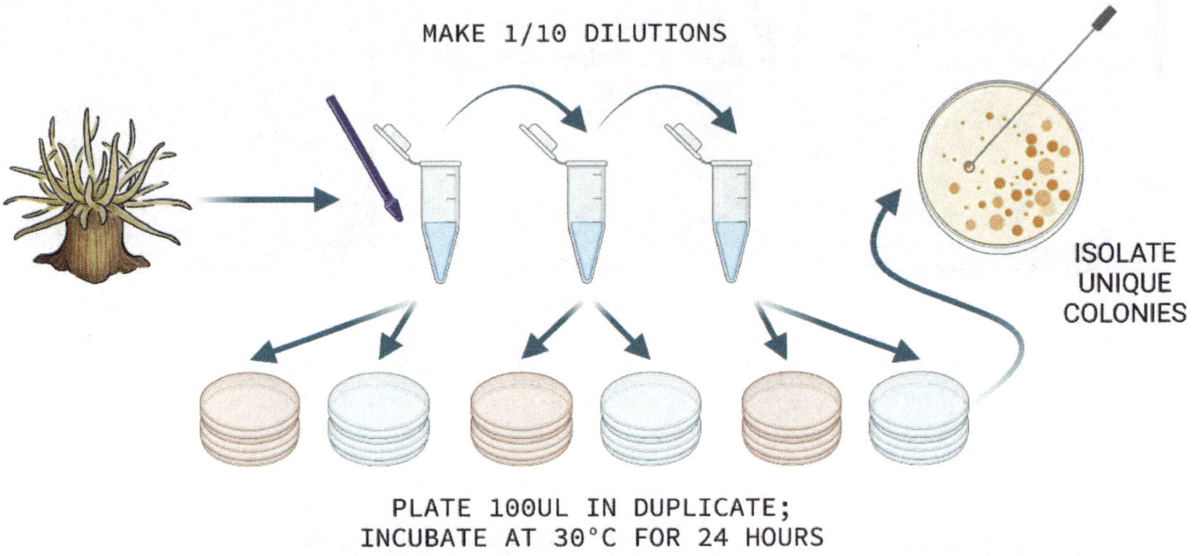

Figure 1. Process of bacterial isolation from Diadumene lineata and Metridium senile. Although only D. lineata is shown in this diagram, the same procedure was performed for both anemone species. The brown plates represent the anemone medium, while the blue plates represent marine agar. Created with BioRender.com.

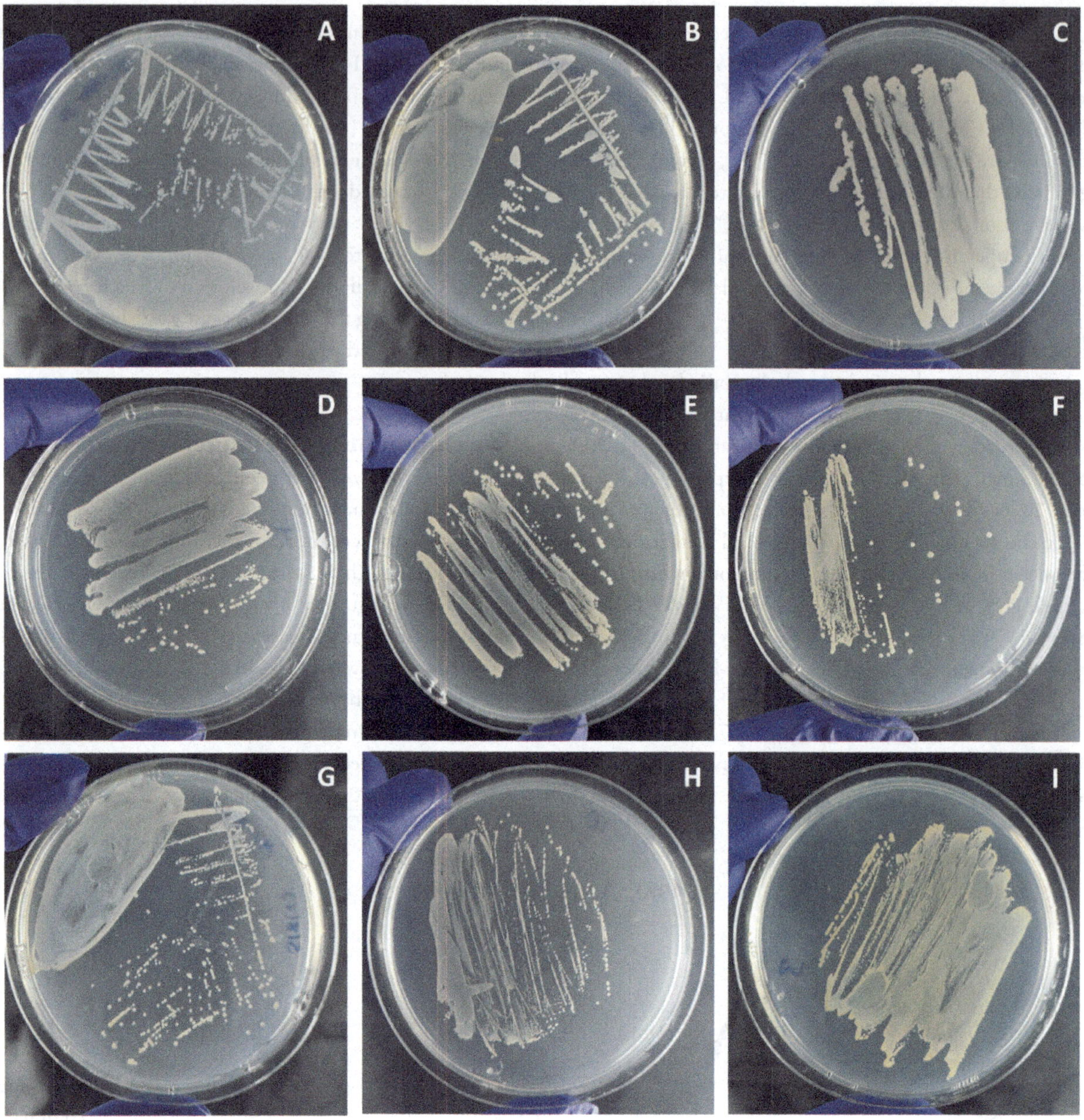

Figure 2. Subset of the anemone-associated bacteria isolates that were selected for the thermal stress assay. (A) DL-AM-10 Pseudoalteromonas, (B) MS-AM-01 Pseudoalteromonas, (C) MS-MA-06 Pseudoalteromonas, (D) DL-AM-06 Peribacillus, (E) DL-MA-07 Vibrio/Peribacillus, (F) MS-MA-02 Peribacillus, (G) MS-MA-01 Litoreibacter, (H) MS-MA-05 Colwellia, and (I) DL-MA-08 Shewanella. There was a mix-up and DL-MA-07 needs to be re-sequenced to determine if it is Vibrio or Peribacillus.

was prepared as both a broth and in solid agar plates, with the solid medium requiring the addition of 20 g of agar per 1000 mL. Marine agar was prepared for comparison, following the product instructions for BD Difco Marine Broth (product #2216) and adding 15 g/mL of agar (the same concentration as BD Difco Marine Agar).

Isolation of Bacteria

Two small anemones (one *D. lineata* and one *M. senile*) were rinsed with sterile deionized water and then transferred to a 1.5 mL microcentrifuge tube filled with 1000 mL of 35 ppt artificial seawater. They were macerated using a sterile plastic pestle and then vortexed vigorously. 1:10 serial dilutions were prepared from each macerated anemone using 100 uL samples in 900 uL of 35 ppt artificial seawater. 100 uL from the 10^0, 10^{-1}, and 10^{-2} dilutions were plated on both anemone agar and marine agar in duplicate. Negative controls for both media were prepared using 100 uL of sterile 35 ppt artificial seawater. The plates were incubated for approximately 48 hours at 30°C, which is on the high end of temperature tolerance for *Diadumene lineata*. While a longer incubation period of one week was initially planned to account for slower growing bacteria, the plates became too overgrown to effectively isolate individual colonies. Colonies exhibiting a variety of morphologies were streaked on fresh anemone or marine medium. Isolates recovered on anemone medium were also plated on marine agar to see if they could grow without the presence of anemone tissue (***Figure 1***).

Cryopreservation

All isolates were cryopreserved at -80°C using 50% v/v glycerol. Cultures were grown in 10 mL of either marine or anemone broth for approximately 48 hours, along with 10 mL of each broth as a negative control. 500 uL of broth was combined with 500 uL of sterile 50% v/v glycerol in Nalgene Cryogenic Tubes (cat. #5000-1012). Both the 48-hour cultures and the frozen isolates were plated on their respective media and incubated at 30°C for 24 hours to verify no contamination was introduced at any step of the cryopreservation process.

Sanger Sequencing

Single colony PCR was performed on fresh 24-hour cultures using Thermo DreamTaq Green PCR Master Mix (cat. #K1081). The 27F (5'-AGAGTTTGATCCT-GGCTCAG-3') and 1492R (5'-GGTTACCTTGTTAC-GACTT-3') universal bacteria primers were used to target the 16S rRNA gene for identification. PCR reactions were 50 uL total, including the addition of 0.5 uL dimethyl sulfoxide to encourage lysis. PCR reaction: 95°C for 4:00 minutes, followed by 35 cycles of 94°C for 1:00 minute, 68°C for 1:00 minute, and 72°C for 1:00 minute, finishing with 72°C for 7:00 minutes before holding at 4°C indefinitely. The PCR products were checked for successful amplification using gel electrophoresis and then purified using the PureLink kit (cat. #K310002). A few modifications were made to the manufacturer purification protocol to increase the concentration of DNA, including increasing incubation times by two to three minutes and eluting the DNA in 25 uL instead of 50 uL. DNA concentrations were measured using a ND-1000 Nanodrop, and then shipped to the Eurofins Sanger Sequencing service. Sequences were trimmed and checked for errors using Applied Biosystems Sequence Scanner 2, then identified to genus using NCBI BLAST.

Isolate Thermal Stress Assay

An initial trial was conducted to make sure the marine broth did not interfere with measuring the optical density. For the trial, a single isolate (DL-MA-07) was grown over the course of six hours, measuring growth at hours three through six by both optical density and plating on marine agar. Uneven spreading made some of the plates difficult to analyze, so DNA from frozen samples (at each time point) were extracted using the ZymoBiomics Miniprep Kit and qPCR was run on the 16S

Genus	All Isolates (N=33)	Marine Agar (N=16)	Anemone Medium (N=17)	D. lineata (N=19)	M. senile (N=14)
Pseudoalteromonas	51.52% (17/33)	68.75% (11/16)	64.71% (11/17)	57.89% (11/19)	42.86% (6/14)
Peribacillus	27.27% (9/33)	25.00% (4/16)	29.41% (5/17)	21.05% (4/19)	35.71% (5/14)
Arcobacter	3.03% (1/33)	6.26% (1/16)	0	5.26% (1/19)	0
Shewanella	3.03% (1/33)	6.26% (1/16)	0	5.26% (1/19)	0
Colwellia	6.06% (2/33)	6.26% (1/16)	5.88% (1/17)	5.26% (1/19)	7.14% (1/14)
Vibrio	3.03% (1/33)	6.26% (1/16)	0	5.26% (1/19)	0
Litoreibacter	3.03% (1/33)	6.26% (1/16)	0	0	7.14% (1/14)
Macrococcus	3.03% (1/33)	6.26% (1/16)	0	0	7.14% (1/14)
Total Richness	**8**	**8**	**3**	**6**	**5**

Table 1. Summary of 33 bacteria isolated and identified to genus using sanger sequencing and NCBI BLAST. The first column shows the percentage of each genus isolated out of the total number of isolates. The second column breaks down the isolated by which medium they were cultured on, while the third column breaks down the same isolates by which anemone species they came from.

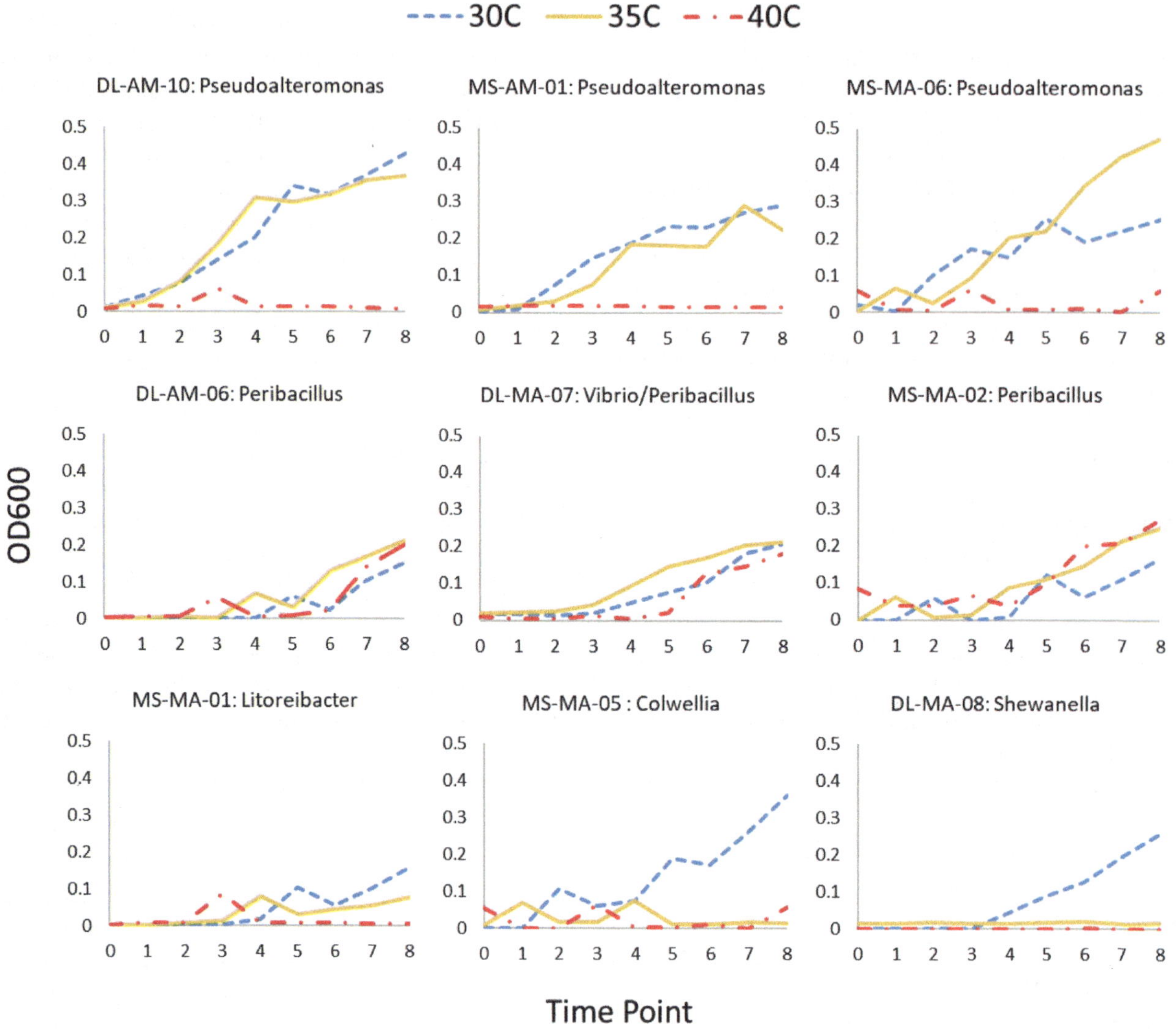

Figure 3. Growth of nine isolated bacteria from the sea anemones Diadumene lineata and Metridium senile. The names include both the isolate ID and the assigned genus. The first portion of the isolate ID denotes anemone species (DL=D. lineata, MS=M. senile), the second portion indicates growth medium (AM=anemone medium, MA=marine agar) and then last portion is the isolate number. Growth was measured at regular intervals over the course of 9 hours using optical density at a wavelength of 600.

rRNA gene to estimate the change in cell abundance over time. The OD600 turbidity measurements and qPCR results were close enough to continue with the experiment (only one time point was lower than expected due to a low yield of extracted DNA) (***Figure 4, Supplemental***).

For the thermal stress assay, 10 isolates were grown for 24 hours in 5 mL of marine broth at 30°C on an orbital shaker. Sterile marine broth was used as a negative control at all stages of the assay. A wide representation of genera from both anemones were selected for the assay, with the exclusion of an *Arcobacter sp.* which failed to grow from the -80°C stock (and was replaced by an additional *Pseudoalteromonas sp.).* Three replicate one mL aliquots were cryopreserved at -20°C with one mL of sterile 50% v/v glycerol for each isolate, so that each of the three temperature treatments would start with a similar concentration of bacteria (per isolate, all isolates were not diluted to the same concentration). Isolates were plated on marine agar at the time of cryopreservation to ensure no contamination was introduced during culturing (***Figure 2***). Each one mL aliquot was used to inoculate 50 mL of marine broth, and the flasks were placed on an incubating orbital shaker to ensure the cultures stayed homogenized and aerated during growth. Nine time points were sampled over the course of approximately nine hours. At each sampling time point, 100 uL samples were transferred in triplicate to a 96-well plate after every hour of growth and the OD600 was measured using the SpectraMax i3 plate reader. There were three temperature treatments total: 30°C, 35°C, and 40°C. Each treatment was run on a separate day so that the same incubating orbital shaker could be used, and a HOBO datalogger was placed in the incubating orbital shaker to track any fluctuations in temperature.

Before analysis, one isolate (MS-MA-04) was excluded due to poor recovery from cryopreservation; although it grew in the initial culturing step from the -80°C freezer it did not grow substantially in any temperature treatment. The absorbance values of all replicates were averaged, then the average of the negative control replicates was subtracted. Very small negative values that occurred when measuring optical densities close to zero were converted to zero for plotting growth over time.

Results

Thirty-three bacteria were isolated in total: 19 from *D. lineata* and 14 from *M. senile*. From *D. lineata*, eight were from marine agar and 11 were from anemone medium; from *M.senile* eight were from marine agar and six were from anemone medium. All bacteria originally isolated on anemone medium also grew on marine agar without issue. Eight unique genera were recovered on marine agar, while only three were recovered on anemone medium. Although most of the bacteria recovered on both media were *Pseudoalteromonas spp.* or *Peribacillus spp.*, they were heavily selected for on the anemone medium and the only other genus isolated was *Colwellia* (***Table 1***). A similar number of genera were recovered from both *D. lineata* and *M. Senile*, however different unique taxa were represented.

Isolates of *Pseudoalteromonas* exhibited robust growth at both the 30°C and 35°C temperatures, with very little growth at 40°C (***Figure 3***). In particular, one *Pseudoalateromonas* isolate from *M. senile* showed a strong preference for the 35°C treatment. Isolates of *Peribacillus* had the broadest range of temperature tolerance, following a similar growth pattern from 30-40°C. In contrast, *Litoreibacter*, *Colwellia*, and *Shewanella* isolates grew best in the 30°C treatment, with minimal growth at 35°C (for *Litoreibacter* only) and no growth at 40°C. HOBO dataloggers reported average experiment temperatures of 30.40°C (st.deviation of 0.14°C), 35.61°C (st.deviation of 0.28°C), and 40.95°C (st.deviation of 0.37°C).

Discussion

The anemone medium captured a lower diversity of microorganisms than the marine agar and heavily selected for bacteria from the genera *Pseudoalteromonas* and *Peribacillus*. Lin et al., (2018) found that *Pseudoalteromonas* was also a dominant genus when isolating bacteria from kelp on medium where alginate was the sole source of nutrients. Although the diversity was higher on marine agar, it may preferentially select for *Gammaproteobacteria* (Joint et al., 2010) which include the genera *Pseudoalteromonas, Shewanella, Vibrio*, and *Colwellia* from this set of isolates. When comparing our isolated taxa to the results of 16S rRNA amplicon sequencing from the same anemones species (Lund, 2023), only 3.6% of genera were recovered on marine agar and 1.6% on anemone medium, however these anemones were living in the laboratory for multiple months prior to sequencing, which could have reduced the bacterial diversity.

In the thermal stress assay, *Peribacillus* isolates exhibited the broadest temperature tolerance between 30°C and 40°C, however *Pseudoalteromonas* isolates also grew well up to 35°C. In contrast, *Litoreibacter*, *Colwellia*, and *Shewanella* isoaltes all preferred lower temperatures and grew the best at 30°C. *Litoreibacter spp.* have solely been isolated from marine environments and are often psychrotolerant, with optimal growth temperatures ranging from 25°C-30°C (Kanamuro et al., 2021; Kim et al., 2014; Romanenko et al., 2011). A dark diffuse pigment was observed in our isolate (***Figure 5, Supplemental***), which has similarly been observed in other *Litoreibacter sp.* cultured from marine sediment (Romanenko et al., 2011). This may be a pyomelanin-like compound, which is associated with protecting cells during heat stress in other bacterial taxa (Zeng et al., 2017) and warrants further analysis. Similarly, *Colwellia sp.* also tend to be psychrotolerant or psychrophilic (Bowman et al., 1998; Deming & Junge, 2015) and although species of *Shewanella* are known pathogens in humans and aquatic organisms, with infections typically occurring in warm climates, the majority of species in this genus are also psychrophilic (Holt et al., 2005). This suggests the *Colwellia* and *Shewanella* isolates in this study are particularly heat tolerant in comparison with other members of their genera, since 30°C is still considered very warm for marine environments. While *Pseudoalteromonas spp.* may be the best candidate for beneficial host-microbe interactions, based on its thermotolerance aligning with *D. lineata*'s upper temperature limit, the possible pyomelanin production in *Literoreibacter* is also worth exploring.

The capacity for these genera to feed on anemone tissue, in conjunction with their high temperature tolerance, may suggest the potential for becoming opportunistic pathogens.

Both genera were abundant in samples from both anemone species, and exhibited similar growth patterns regardless of which anemone they were cultured from or host temperature tolerance. In the onion-based OEM medium, known pathogens grow better on onion tissue, while non-pathogenic onion-associated bacteria grow better on less selective LB medium (Zaid et al., 2012). However, in the 16S rRNA data from a temperature gradient experiment on *D. lineata* and *M. senile* we do not see increases in either of these genera in the mortality community, and *Pseudoalteromonas spp.* are generally prevalent in *M.senile* (Lund, 2023). It may be that these genera are feeding on the mucus of the anemone and not directly on the tissue. *Pseudoalteromonas spp.* have been found in the mucus layer of corals, some of which had antibacterial properties that could protect the host (Shnit-Orland et al., 2012). Future research could either culture isolates from different parts of the anemone, or test if the isolates in this study grow on a medium prepared only with anemone mucus versus anemone tissue.

Alternatively, both *Pseudoalteromonas* and *Peribacillus* have the potential to combat opportunistic bacteria from the environment when the host is experiencing heat stress. While less research has investigated *Peribacillus spp.* in marine hosts, hemolytic *Pseudoalteromonas spp.* have been found living aggregated inside the tentacles of the anemone *Sargatia elegans* (Schuett et al., 2007) and *Pseudoalteromonas* has been identified as a putative beneficial genus for corals (Rosado et al., 2019). Not only have *Pseudoalteromonas sp.* have been noted for their capacity to readily colonize surfaces through the rapid formation of biofilms (Wu et al., 2021) they are also prolific producer of bioactive secondary metabolites (Atencio et al., 2020). Similarly, marine *Bacillus* also produce a variety of secondary metabolites of interest (Mondol et al., 2013), including lipopeptides with surfactant properties that could be used in oil spill remediation (Hentati et al., 2019), antibiotics, and antifungals (Franks et al., 2006; Shnit-Orland et al., 2012).

The highly selective nature of the anemone medium could benefit research in marine probiotics and bioremediation through targeting novel *Pseudoalteromonas spp.* and *Peribacillus spp.* One issue to consider when using this medium in future research is the variation that occurs between batches of ground anemone. The anemone tissue had varying amounts of air whipped into it during the grinding process, resulting in some 50 mL aliquots containing more air pockets than others. There were also differences in viscosity between batches, likely related to the quantity of mucus, which affected how well the anemone tissue homogenized in the medium. Although this never prevented the growth of the cultured isolates, it is an important consideration for reproducibility in experiments. A final potential issue is that the anemone medium had a very short shelf life (even when refrigerated) and lasted about five days before bacteria stopped growing on it entirely. Throughout this study, the medium was prepared fresh the day before it was intended to be used, which may be inconvenient for some experimental designs.

Microbiological medium made from anemone tissue successfully cultured microbiota and was more selective than commercially available marine agar. Similar dominant genera (*Pseudoalteromonas* and *Peribacillus*) were cultured from both estuarine anemone species *D. lineata* and *M. senile* on both media (but were preferentially selected for by the anemone medium). Although we were unable to detect a difference between bacteria recovered from two different anemone species, the robust growth at 30-35°C in conjunction with research on probiotics in corals (Rosado et al., 2019) highlights the genus *Pseudoalteromonas* as a potential candidate for beneficial anemone-microbe interactions under high temperature stress. Further research should explore the production of secondary metabolites in these isolates, particularly antimicrobial/antifungal activity towards environmental microbiota. This work supports culturing as a critical component of unraveling the individual functions of these bacteria in anemone microbial communities.

Citations

Atencio, L. A., Boya P., C. A., Martin H., C., Mejía, L. C., Dorrestein, P. C., & Gutiérrez, M. (2020). Genome Mining, Microbial Interactions, and Molecular Networking Reveals New Dibromoalterochromides from Strains of Pseudoalteromonas of Coiba National Park-Panama. *Marine Drugs, 18*(9), Article 9. https://doi.org/10.3390/md18090456

Bowman, J. P. (2007). Bioactive compound synthetic capacity and ecological significance of marine bacterial genus pseudoalteromonas. *Marine Drugs, 5*(4), 220–241. https://doi.org/10.3390/md504220

Bowman, J. P., GOSINK, J. J., McCAMMON, S. A., LEWIS, T. E., NICHOLS, D. S., NICHOLS, P. D., SKERRATT, J. H., STALEY, J. T., & McMEEKIN, T. A. (1998). Colwellia demingiae sp. nov., Colwellia hornerae sp. nov., Colwellia rossensis sp. nov. and Colwellia psychrotropica sp. nov.: Psychrophilic Antarctic species with the ability to synthesize docosahexaenoic acid (22:6ω3). *International Journal of Systematic and Evolutionary Microbiology, 48*(4), 1171–1180. https://doi.org/10.1099/00207713-48-4-1171

Chen, H., Nwe, P.-K., Yang, Y., Rosen, C. E., Bielecka, A. A., Kuchroo, M., Cline, G. W., Kruse, A. C., Ring, A.

M., Crawford, J. M., & Palm, N. W. (2019). A Forward Chemical Genetic Screen Reveals Gut Microbiota Metabolites That Modulate Host Physiology. *Cell, 177*(5), 1217-1231.e18. https://doi.org/10.1016/j.cell.2019.03.036

Cunning, R., & Baker, A. C. (2020). Thermotolerant coral symbionts modulate heat stress-responsive genes in their hosts. *Molecular Ecology, 29*(15), 2940–2950. https://doi.org/10.1111/mec.15526

Cziesielski, M. J., Liew, Y. J., Cui, G., Schmidt-Roach, S., Campana, S., Marondedze, C., & Aranda, M. (2018). Multi-omics analysis of thermal stress response in a zooxanthellate cnidarian reveals the importance of associating with thermotolerant symbionts. *Proceedings of the Royal Society B: Biological Sciences, 285*(1877). https://doi.org/10.1098/rspb.2017.2654

Debbab, A., Aly, A. H., Lin, W. H., & Proksch, P. (2010). Bioactive Compounds from Marine Bacteria and Fungi. *Microbial Biotechnology, 3*(5), 544–563. https://doi.org/10.1111/j.1751-7915.2010.00179.x

Deming, J. W., & Junge, K. (2015). Colwellia. In *Bergey's Manual of Systematics of Archaea and Bacteria* (pp. 1–12). John Wiley & Sons, Ltd. https://doi.org/10.1002/9781118960608.gbm01090

Fautin, D. G., & Hand, C. (2007). Anthozoa. In *The Light and Smith Manual: Intertidal inver-tebrates from central California to Oregon.* (pp. 173–184). University of California Press.

Fautin, D. G., Siebert, A. E., & Kozloff, E. N. (1987). Class Anthozoa. In *Marine Invertebrates of the Pacific Northwest* (pp. 68–78). University of Washington Press.

Franks, A., Egan, S., Holmström, C., James, S., Lappin-Scott, H., & Kjelleberg, S. (2006). Inhibition of Fungal Colonization by Pseudoalteromonas tunicata Provides a Competitive Advantage during Surface Colonization. *Applied and Environmental Microbiology, 72*(9), 6079–6087. https://doi.org/10.1128/AEM.00559-06

Glon, H., Haruka, Y., Daly, M., & Nakaoka, M. (2019). Temperature and salinity survival limits of the fluffy sea anemone, Metridium senile (L.), in Japan. *Hydrobiologia, 830*(1), 303–315. https://doi.org/10.1007/s10750-018-3879-2

Hayatsu, M., Tago, K., & Saito, M. (2008). Various players in the nitrogen cycle: Diversity and functions of the microorganisms involved in nitrification and denitrification. *Soil Science and Plant Nutrition, 54*(1), 33–45. https://doi.org/10.1111/j.1747-0765.2007.00195.x

Hentati, D., Chebbi, A., Hadrich, F., Frikha, I., Rabanal, F., Sayadi, S., Manresa, A., & Chamkha, M. (2019). Production, characterization and biotechnological potential of lipopeptide biosurfactants from a novel marine Bacillus stratosphericus strain FLU5. *Ecotoxicology and Environmental Safety, 167*, 441–449. https://doi.org/10.1016/j.ecoenv.2018.10.036

Holt, H. M., Gahrn-Hansen, B., & Bruun, B. (2005). Shewanella algae and Shewanella putrefaciens: Clinical and microbiological characteristics. *Clinical Microbiology and Infection, 11*(5), 347–352. https://doi.org/10.1111/j.1469-0691.2005.01108.x

Hu, J., Wei, Z., Friman, V.-P., Gu, S., Wang, X., Eisenhauer, N., Yang, T., Ma, J., Shen, Q., Xu, Y., & Jousset, A. (2016). Probiotic Diversity Enhances Rhizosphere Microbiome Function and Plant Disease Suppression. *MBio, 7*(6), e01790-16. https://doi.org/10.1128/mBio.01790-16

Joint, I., Mühling, M., & Querellou, J. (2010). Culturing marine bacteria – an essential prerequisite for biodiscovery. *Microbial Biotechnology, 3*(5), 564–575. https://doi.org/10.1111/j.1751-7915.2010.00188.x

Kanamuro, M., Sato-Takabe, Y., Muramatsu, S., Hirose, S., Muramatsu, Y., Takaichi, S., & Hanada, S. (2021). Litoreibacter roseus sp. Nov., a novel bacteriochlorophyll a-containing bacterium. *International Journal of Systematic and Evolutionary Microbiology, 71*(3), 004679. https://doi.org/10.1099/ijsem.0.004679

Kehe, J., Kulesa, A., Ortiz, A., Ackerman, C. M., Thakku, S. G., Sellers, D., Kuehn, S., Gore, J., Friedman, J., & Blainey, P. C. (2019). Massively parallel screening of synthetic microbial communities. *Proceedings of the National Academy of Sciences, 116*(26), 12804–12809. https://doi.org/10.1073/pnas.1900102116

Kim, Y.-O., Park, S., Nam, B.-H., Park, J.-M., Kim, D.-G., & Yoon, J.-H. (2014). Litoreibacterascidiaceicola sp. Nov., isolated from the golden sea squirt Halocynthiaaurantium. *International Journal of Systematic and Evolutionary Microbiology, 64*(Pt_8), 2545–2550. https://doi.org/10.1099/ijs.0.064196-0

Lawley, T. D., Clare, S., Walker, A. W., Stares, M. D., Connor, T. R., Raisen, C., Goulding, D., Rad, R., Schreiber, F., Brandt, C., Deakin, L. J., Pickard, D. J., Duncan, S. H., Flint, H. J., Clark, T. G., Parkhill, J., & Dougan, G. (2012). Targeted restoration of the intestinal microbiota with a simple, defined bacteriotherapy resolves relapsing Clostridium difficile disease in mice. *PLoS Pathogens, 8*(10), e1002995. https://doi.org/10.1371/journal.ppat.1002995

Lin, J. D., Lemay, M. A., & Parfrey, L. W. (2018). Diverse Bacteria Utilize Alginate Within the Microbiome of the Giant Kelp Macrocystis pyrifera. *Frontiers in Microbiology, 9*. https://www.frontiersin.org/articles/10.3389/fmicb.2018.01914

Locey, K. J., & Lennon, J. T. (2016). Scaling laws predict global microbial diversity. *Proceedings of the National Academy of Sciences, 113*(21), 5970–5975. https://doi.org/10.1073/pnas.1521291113

Lund, P. K. (2023). *Host-microbe interactions in non-native estuarine anemones: Biogeography and temperature* [Unpublished Masters Thesis]. California State Polytechnic University, Humboldt.

McAnulty, S. J., Kerwin, A. H., Koch, E., Nuttall, B., Suria, A. M., Collins, A. J., Schleicher, T. R., Rader, B. A., & Nyholm, S. V. (2023). "Failure To Launch": Development of a Reproductive Organ Linked to Symbiotic Bacteria. *MBio, 0*(0), e02131-22. https://doi.org/10.1128/mbio.02131-22

Mondol, M. A. M., Shin, H. J., & Islam, M. T. (2013). Diversity of Secondary Metabolites from Marine Bacillus Species: Chemistry and Biological Activity. *Marine Drugs, 11*(8), Article 8. https://doi.org/10.3390/md11082846

Muriel-Millán, L. F., Millán-López, S., & Pardo-López, L. (2021). Biotechnological applications of marine bacteria in bioremediation of environments polluted with hydrocarbons and plastics. *Applied Microbiology and Biotechnology, 105*(19), 7171–7185. https://doi.org/10.1007/s00253-021-11569-4

Parfrey, L. W., Moreau, C. S., & Russell, J. A. (2018). Introduction: The host-associated microbiome: Pattern, process and function. *Molecular Ecology, 27*(8), 1749–1765. https://doi.org/10.1111/mec.14706

Raina, J.-B., Tapiolas, D., Willis, B. L., & Bourne, D. G. (2009). Coral-Associated Bacteria and Their Role in the Biogeochemical Cycling of Sulfur. *Applied and Environmental Microbiology, 75*(11), 3492–3501. https://doi.org/10.1128/AEM.02567-08

Rathgeber, C., Yurkova, N., Stackebrandt, E., Schumann, P., Humphrey, E., Beatty, J. T., & Yurkov, V. (2006). Metalloid Reducing Bacteria Isolated from Deep Ocean Hydrothermal Vents of the Juan de Fuca Ridge, Pseudoalteromonas telluritireducens sp. Nov. And Pseudoalteromonas spiralis sp. Nov. *Current Microbiology, 53*(5), 449–456. https://doi.org/10.1007/s00284-006-0320-2

Romanenko, L. A., Tanaka, N., Frolova, G. M., Svetashev, V. I., & Mikhailov, V. V. (2011). Litoreibacter albidus gen. Nov., sp. Nov. And Litoreibacter janthinus sp. Nov., members of the class Alphaproteobacteria isolated from the seashore. *International Journal of Systematic and Evolutionary Microbiology, 61*(1), 148–154. https://doi.org/10.1099/ijs.0.019513-0

Rosado, P. M., Leite, D. C. A., Duarte, G. A. S., Chaloub, R. M., Jospin, G., Nunes da Rocha, U., P. Saraiva, J., Dini-Andreote, F., Eisen, J. A., Bourne, D. G., & Peixoto, R. S. (2019). Marine probiotics: Increasing coral resistance to bleaching through microbiome manipulation. *The ISME Journal, 13*(4), Article 4. https://doi.org/10.1038/s41396-018-0323-6

Ryan, W., & Kubota, S. (2016). Morphotype distribution of the sea anemone Diadumene lineata in Tanabe Bay, Wakayama: A comparison with Uchida (1936) after 80 years. *Publications of the Seto Marine Biological Laboratory, 44*, 1–6. https://doi.org/10.5134/209001

Santoro, E., Borges, R., Espinoza, J., Freire, M., Messias, C., Villela, H., Pereira, L., Vilela, C., Rosado, J., Cardoso, P., Rosado, P., Assis, J., Duarte, G., Perna, G., Rosado, A., Macrae, A., Dupont, C., Nelson, K., Sweet, M., & Peixoto, R. (2021). Coral microbiome manipulation elicits metabolic and genetic restructuring to mitigate heat stress and evade mortality. *Science Advances, 7*. https://doi.org/10.1126/sciadv.abg3088

Schuett, C., Doepke, H., Grathoff, A., & Gedde, M. (2007). Bacterial aggregates in the tentacles of the sea anemone Metridium senile. *Helgoland Marine Research, 61*(3), 211–216. https://doi.org/10.1007/s10152-007-0069-4

Shick, J. M. (1976). Ecological Physiology and Genetics of the Colonizing Actinian Haliplanella Luciae. In G. O. Mackie (Ed.), *Coelenterate Ecology and Behavior* (pp. 137–146). Springer US. https://doi.org/10.1007/978-1-4757-9724-4_15

Shnit-Orland, M., Sivan, A., & Kushmaro, A. (2012). Antibacterial Activity of Pseudoalteromonas in the Coral Holobiont. *Microbial Ecology, 64*(4), 851–859. https://doi.org/10.1007/s00248-012-0086-y

Stock, W., Blommaert, L., De Troch, M., Mangelinckx, S., Willems, A., Vyverman, W., & Sabbe, K. (2019). Host specificity in diatom–bacteria interactions alleviates antagonistic effects. *FEMS Microbiology Ecology, 95*(11), fiz171. https://doi.org/10.1093/femsec/fiz171

Tortell, P. D., Maldonado, M. T., Granger, J., & Price, N. M. (1999). Marine bacteria and biogeochemical cycling of iron in the oceans. *FEMS Microbiology Ecology, 29*(1), 1–11. https://doi.org/10.1111/j.1574-6941.1999.tb00593.x

Uchida, T. (1932). Occurrence in Japan of Diadumene luciae, a remarkable Actinian of rapid dispersal. *Journal of Faculty of Science, Hokkaido Imperial University, 2*(2), 18.

Vázquez, J. A., González, M. P., & Murado, M. A. (2004). A new marine medium: Use of different fish peptones and comparative study of the growth of selected species of marine bacteria. *Enzyme and Microbial Technology, 35*(5), 385–392. https://doi.org/10.1016/j.enzmictec.2004.02.007

Wu, Z., Wu, Y., Huang, Y., He, J., Su, P., & Feng, D. (2021). Insights into the planktonic to sessile transition in a marine biofilm-forming Pseudoalteromonas isolate using comparative proteomic analysis. *Aquatic Microbial Ecology, 86*, 69–84. https://doi.org/10.3354/ame01959

Zaid, A. M., Bonasera, J. M., & Beer, S. V. (2012). OEM—A new medium for rapid isolation of onion-pathogenic and onion-associated bacteria. *Journal of Microbiological Methods, 91*(3), 520–526. https://doi.org/10.1016/j.mimet.2012.09.031

Zeng, Z., Cai, X., Wang, P., Guo, Y., Liu, X., Li, B., & Wang, X. (2017). Biofilm Formation and Heat Stress Induce Pyomelanin Production in Deep-Sea Pseudoalteromonas sp. SM9913. *Frontiers in Microbiology, 8*. https://www.frontiersin.org/articles/10.3389/fmicb.2017.01822

Supplementary Information

DL-MA-05 came up as *Poseidonibacter* on BLAST, but has been labeled *Arcobacter* based on recent taxonomic revisions (On et al., 2021). DL-MA-07 is marked with a * because it was originally *Vibrio*, but may have gotten mixed up with a *Peribacillus* isolate. There's no way to know for certainty if the isolate in the heat stress assay was *Vibrio* or *Peribacillus* without re-sequencing the stock from the -80°C freezer. DL-AM-11 is marked with an * because it appears to have been contaminated at one point with a different bacteria (possibly *Stentrophomonas*, but the re-sequencing results were too messy to say for sure).

Before running the heat stress assay on all isolates, a small six-hour trial was performed to make sure the marine broth wasn't interfering with the optical density readings. Only measurements for hours three through were measured, based on the growth patterns of previous trials. In the initial trial of optical density versus plating, the plate results were difficult to count due to high-density splotches of bacteria (particularly at the later time points). During this trial, 900 uL of culture was frozen at -20°C for each time point to use for qPCR in case the plating results were messy. The frozen culture was extracted using the ZymoBiomics Miniprep Kit and qPCR was run using the primers EUB338 (5'-ACTCCTACGGGAGGCAGCAG-3') and EUB518 (5'-ATTACCGCGGCTGCTGG-3'). The results of qPCR follow a similar pattern of growth to optical density, except for time point six. This is likely due to an issue in the extraction process, since time point five had a higher DNA concentration than time point six (8.5 versus 6.5 ng/uL). The results for OD600 versus qPCR were clear enough for me to continue with the heat stress assay on the isolates.

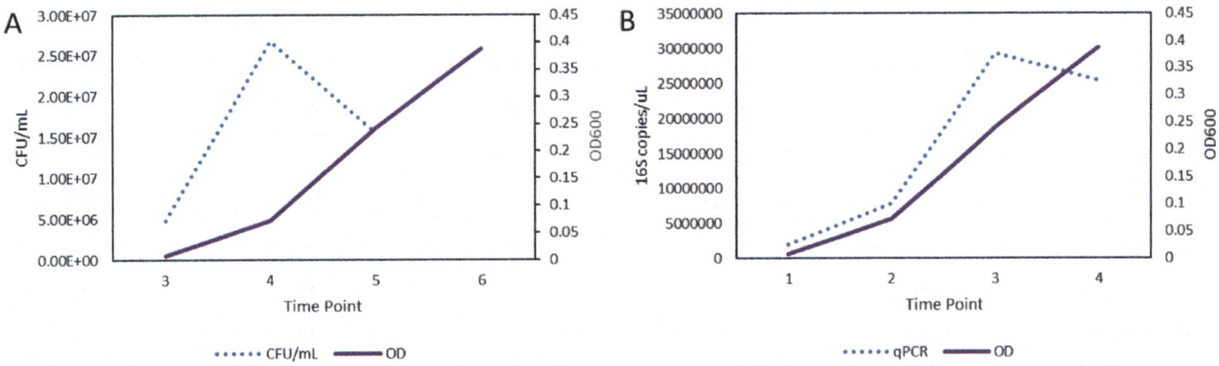

Figure 4. Comparison of measurements for culture growth from hour 3 through hour 6 (for isolate DL-MA-07). Optical density at wavelength 600 was compared with (A) plated CFU/mL and (B) 16S rRNA gene copies per uL of DNA. Samples from the same culture were used for all measurements.

Isolate ID	Genus	Quality	BP	% Identity	Accession #
DL-MA-01	Pseudoalteromonas	Good	999	99.70%	MN889141.1
DL-MA-02	Peribacillus	Good	989	99.90%	MN826155.1
DL-MA-03	Peribacillus	Good	963	99.90%	MN826155.1
DL-MA-04	Pseudoalteromonas	Good	945	100.00%	MH061228.1
DL-MA-05	Arcobacter	Good	1027	99.22%	OP709877.1
DL-MA-06	Pseudoalteromonas	Good	948	100.00%	JX525985.1
DL-MA-07	Vibrio*	Good	956	99.90%	MK720201.1
DL-MA-08	Shewanella	Good	991	100.00%	KR270269.1
DL-AM-01	Pseudoalteromonas	Good	1006	99.60%	MK439566.1
DL-AM-02	Pseudoalteromonas	*Okay*	787	99.24%	MK439566.1
DL-AM-03	Peribacillus	Good	959	99.79%	LT838048.1
DL-AM-04	Pseudoalteromonas	*Bad*	746	99.46%	KP980759.1
DL-AM-05	Pseudoalteromonas	Good	952	99.79%	P118496.1
DL-AM-06	Peribacillus	Good	961	99.90%	LT838048.1
DL-AM-07	Pseudoalteromonas	Good	938	99.89%	MN746121.1
DL-AM-08	Pseudoalteromonas	Good	916	100.00%	MK577380.1
DL-AM-09	Pseudoalteromonas	Good	1002	99.90%	MN889141.1
DL-AM-10	Pseudoalteromonas	Good	1012	100.00%	MN889157.1
DL-AM-11	Colwellia*	*Bad*	871	98.17%	JX436365.1
MS-MA-01	Litoreibacter	Good	1001	100.00%	OL662961.1
MS-MA-02	Peribacillus	Good	894	100.00%	MN826155.1
MS-MA-03	Peribacillus	Good	874	100.00%	OM281797.1
MS-MA-04	Macrococcus	Good	863	99.88%	OK103765.1
MS-MA-05	Colwellia	Good	1034	99.90%	MW580181.1
MS-MA-06	Pseudoalteromonas	Good	876	100.00%	MN889131.1
MS-MA-07	Pseudoalteromonas	Good	939	100.00%	MN889240.1
MS-MA-08	Pseudoalteromonas	Good	864	100.00%	OQ300257.1
MS-AM-01	Pseudoalteromonas	Good	940	100.00%	DQ492734.1
MS-AM-02	Pseudoalteromonas	Good	814	99.88%	OQ300257.1
MS-AM-03	Pseudoalteromonas	Good	984	99.90%	MN209950.1
MS-AM-04	Peribacillus	Good	935	100.00%	LT838048.1
MS-AM-05	Peribacillus	*Bad*	194	100.00%	MT310824.1
MS-AM-06	Peribacillus	*Okay*	750	100.00%	KX881437.1

Table 2. NCBI BLAST results for bacteria isolated from Diadumene lineata (DL) and Metridium senile (MS). MA indicates it was cultured on marine agar, while AM indicates it was cultures on anemone medium. Quality was assigned based on the SeqScanner2 results. A good quality sequence has a long segment of contiguous high quality basepair assignments.

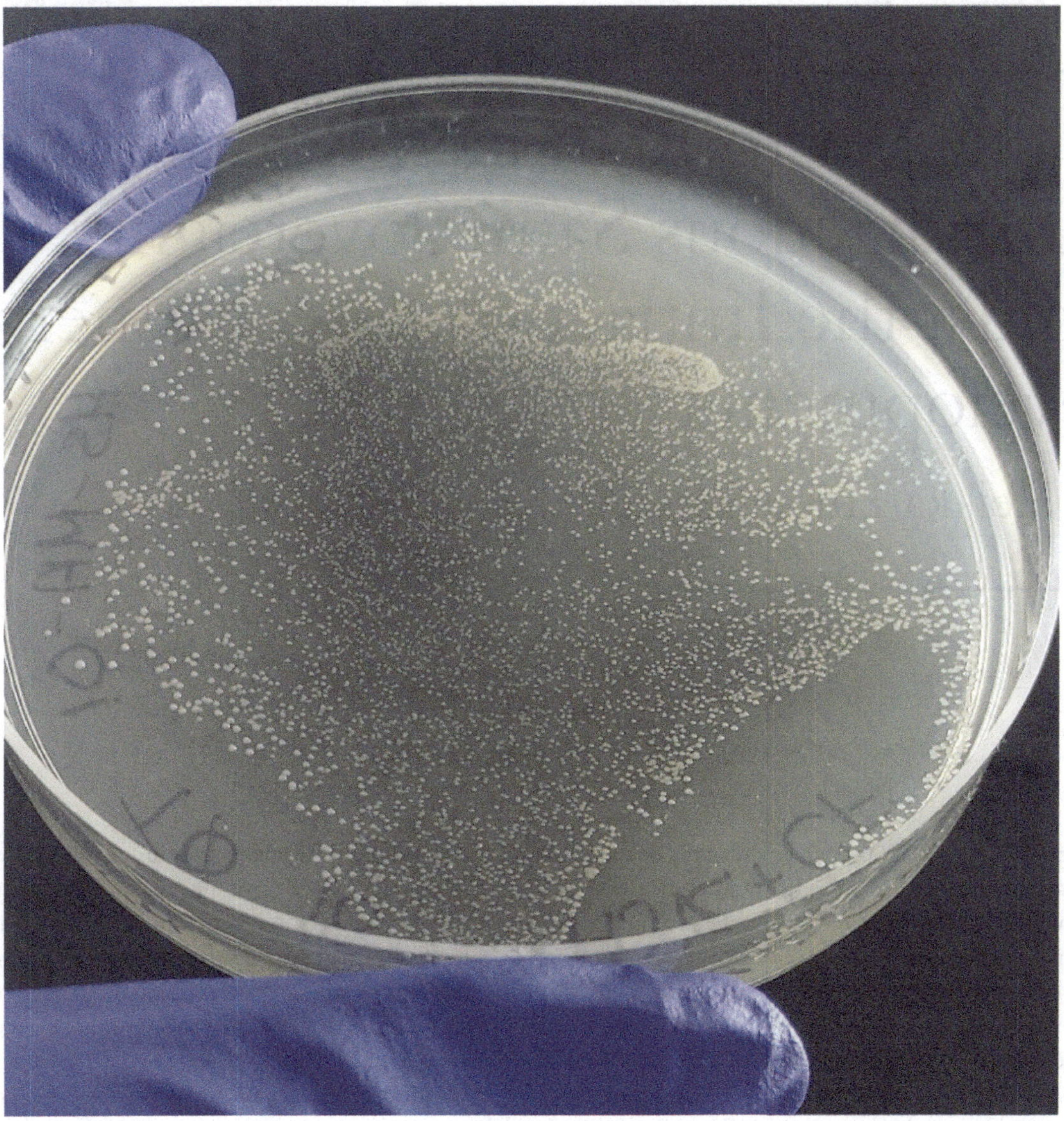

Figure 5. Litoreibacter sp. on marine agar exhibiting the production of a dark diffuse pigment.

IdeaFest
Journal

Challenges to Reindeer, Reciprocity, and Indigenous Sami Sovereignty Amidst the Impact of Green Energy Developments

Lisa Heikka-Huber

Abstract

The Indigenous people of Europe known as the Sami, (also spelled Saami) many of whom live throughout the world, have continued to maintain active nomadic communities today as their ancestors did. A wide spanning region of Northern Europe's Arctic Zone or Sampi often referred to as Fennoscandia, encompasses four countries, Norway, Sweden, Finland, and Russia's Kola Peninsula (Roland & Löffler, 2012). The nomadic Sami people follow the migration pathways of their reindeer herds through the wilderness bi-annually. This paper will discuss many perspectives, including the battle Sami people and other Indigenous communities have endured while combating green energy development from impacting their cultural heritage practice rights, encroachment on their ancestral land, and resources. Starting with defining food sovereignty, then moving on to an introduction of the Sami people and a brief description of their history, along with detailed information of how these generations of Sami people have survived in the Arctic Zone. From there I will discuss specific impacts that climate change and green energy development projects have imposed on reindeer, who are one of the Sami's most important food sources. Thereafter, I will present a detailed summary on a few of the neoliberal development projects in the region which have impacted Sami traditional practices. Including a brief description of wind power generation infrastructures in three Sampi region nation states: Norway, Finland, and Sweden. I will then discuss the importance reindeer hold to Sami foodways, medicinal purposes, and products made from every part of the animal to honor the sacrifice they make for the survival of the Sami culture. The scientific studies of Anna Skarin get to the root of the problem for the Sami: reindeer do not like windmills. The paper also discusses reciprocal relationships between cultural traditions and ecosystem health within Indigenous communities elsewhere. Holding nation states and corporations accountable when violating terms of multiple signed agreements in the name of capital accumulation, green energy developments or other resource extraction that threaten their existence and subsistence practices

Keywords: Food Soveeignty, Indigenous rights, Green energy, Energy development, Windfarms, Wind energy development, Industrialization, Cultural heritage practices, Conservation and green energy, Reindeer herding, Sami people, Sami sovereignty.

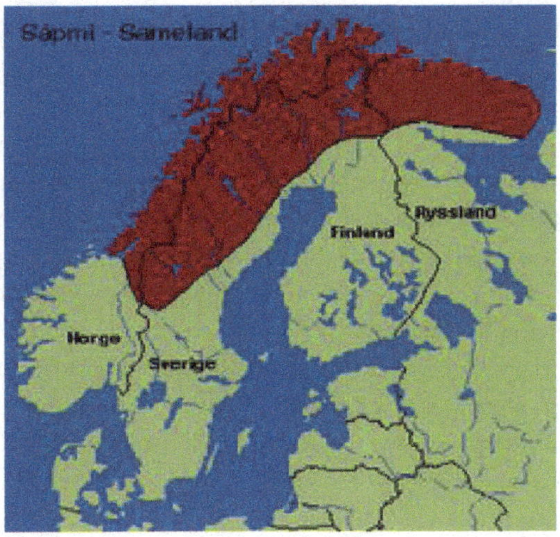

Center for Global Studies. Sampi Map. The Henry M. Jackson School of International Studies: University of Washington.

Ole Martin Wold, Roan Wind Farm. Roan, Fosen, Norway. August 1, 2018

What is Food Sovereignty and Why Does It Matter?

There are four key concepts to understand when considering food sovereignty as stated from the Working Group on Indigenous Food Sovereignty. Those of which include the following: 1) the right to food is sacred 2) daily participation in the food system at all levels is fundamental 3) self-determination is reinforced with access and consumption of culturally appropriate food sources, with rights to quality means of attaining such through hunting, gathering, fishing and other Indigenous methods. Finally, 4) legislation and policy support to help reform current practice and better include Indigenous knowledge systems with colonist laws and regulations that have degraded these activities (McCune, 2021). Having agency over food resources is something all people strive for, but over time has become an ever-widening relationship. In today's modern society it is not easily accessible to have a garden to grow your own food, let alone sustain a whole family diet on foods sourced directly.

Food production in the hands of neoliberal industrialism has perpetuated the stronghold of fossil fuels on the world as population growth yields greater food needs. Industrial agriculture practice predominantly uses fossil fueled machinery to plant and harvest over manual labor, perpetuating the climate crisis, displacing, and dismantling Indigenous communities while taking away their agency of sovereign identity (Morrison, 2011). The battle to protect the rights of Indigenous food sovereignty against colonialism and neoliberal development has been ongoing since colonial settlements began. Though the term sovereignty itself is rooted in European monarchal vocabulary, it is used more commonly when discussing Indigenous rights to self-determination applicable to social, political, and cultural identity, and is nothing without their traditional food sources. Kyle Whyte clarifies the meaning of the "collective capacity" of Indigenous populations when relating to ecological systems that have adapted to larger scale impacts (Whyte, 2016). Reinforcing the importance of "trusted" food sourcing and the sharing of processes. This includes harvesting, seed collections, cultivation, production, and the recirculated refuse within these communities all contributing to the reciprocal relationship between Indigenous people, the land, and their ability to adapt. The ability to feed oneself with agency is what encompasses food sovereignty.

Who are the Sami People?

The Sami people have resided in the boreal region of Sampi in the northern parts of Norway, Sweden, Finland, and the Kola Peninsula of Russia for over 10,000 years (DuBois & Lang, 2013). Living off the land through reciprocal relationships of respect with the region's wilderness and surviving in the harsh climate region for generations. The people of Sampi are known as the Sami, (Levy, 2006). They are the Indigenous people of Europe, unique to each of these nation states, their traditions entwined with reindeer for centuries (Weinstock, 2013). The reindeer survived the region thanks to the Sami. Their cultural connection to the animals, caring for them during periods of over-hunting, eventually starting husbandry practice of the reindeer helped their species repopulate (Demant-Hatt et al., 1908). Fishing the waterways and coastal regions are also traditional cultural practices which directly connect the coastal Sami people to their coastal landscapes (Berg, 2014). The Mountain Sami usually live nomadic lifestyles following reindeer migration pathways to and from seasonal grazing pastures. The coastal Sami live off their fishing practices and regularly trade surplus for other goods with the mountain Sami as well as other cultures (Roland & Löffler, 2012).

As one of the only Indigenous cultures left in Europe, the Sami have battled the impacts left by the industrialization development projects and the encroaching in their regions, including mining,oil drilling, and coupled with more recent green energy infrastructures built directly inside Indigenous migratory pathways, (Skarin et al., 2018). The pristine forests and lands of the most northern parts of Europe are appealing to development corporations looking for wide undisturbed spaces to build out their projects. However, these spaces have been cared for by Sami people for generations and it is the Sami who can be praised for the pristine nature of these lands, no one else. The Sami have Indigenous ancestral rights to these lands as part of their cultural heritage practice. So why do they not hold more authority on how these lands are used by their respective nation states implementing policy to the region (Heaps, 2019)? Despite their presence long before Nordic settler colonists, it is the established governments who still hold the power in determining development prospects, not the Sami.

In 1956 the Sami Council, a voluntary Saami organization was formed to be representative and actively engage with policy related to member organizations from Finland, Sweden, Norway, and Russia's Kola Peninsula (Roland & Löffler, 2012). Organizations such as the Sami Parliament, founded by the Sami to establish their self-determination and sovereignty and collectively take control of their landscapes (Roland & Löffler, 2012). Through varying agreements between nation states, the Sami people have gained wider recognition as sovereign Indigenous people (Heaps, 2019). The concept of sovereignty and the right to self-determination by Indigenous people whose ancestors are rooted to regions long before contact from colonists is recognized by the United Nations as a human right. Indigenous Sami have like many other Indigenous communities a culturally embedded relationship with the land of their people. TEK or Traditional Ecological Knowledge of these lands is immensely valuable information that could be useful today in efforts to help restore degradation impacts of climate change (UN, n.d.). TEK is what sustained their communities for centuries, through practicing and teaching stewardship and cultivation methods that work reciprocally to maintain healthy biodiversity. The same methods used by Sami reindeer herders to maintain their cultural practice of migration with their herds across the Arctic. Their trek from meadow regions between the mountains and coast, led by reindeer foraging for their diets of lichen, moss and other grazed vegetation. In turn the reindeer provide the Sami with necessities they require to sustain the deep temperature shifts of the boreal region. These spaces are pristine, natural environments that Sami people have traditionally cared for as stewards for thousands of years. These ancestral regions are important to their cultural way of life, their heritage is the land. For at least 2,000 years, Sami people have practiced reindeer herding and husbandry (Levy, 2006).

Monarchy, Plague, and the Sami's Pathway to Farm Steading

The Sami way of life likely saved their communities from greater suffering during the Bubonic Plague that hit the region in 1349, known as "Black Death" (Levy, 2006). The Sami people were not as impacted because they could sustain without trade from the outside world, with their diet of reindeer and fish products along with locally foraged grain, berries, and nuts (Levy, 2006). An estimated 60% of Europe succumbed to the disease, leaving empty farmstead lands in Norway that the government encouraged Sami families to take over and use as more sedentary reindeer herding locations (Yngve & Hjalmar, 1965; Roland & Loffler, 2012). Owning land and a farmstead came with some caveats that many Sami were ill suited to handle. Taxation for one, but the taxation of their herds by the head was what the monarchies of Norway required, their way to recoup lost tax revenues from high death rates (Skarin et al, 2015). The land allowed them to establish homes, plant roots enabling food resources to be stored for the long winter months. The Norwegian monarchy introduced a new way of life to the Sami. They still traveled with their herd through migrations, but they were able to return to their farms in between these journeys. With new regulations the Sami were forcibly assimilated into capitalist society they were unaccustomed to, one of many forced assimilations their people would endure. Despite these regulations their culture has sustained; thanks to their reciprocal relationship with the reindeer (Mustonen, 2010).

What is Energy Sovereignty? Is it What it Seems?

In effort to make the evolution of green energy development acceptable to Indigenous communities, developers have worked to normalize the infrastructure needed as ecologically mindful business models that they could do for themselves, in partnership with corporations and nation states (Peart, 2021). By re-branding the idea as energy sovereignty, it is co-optation of the Indigenous people and is only appeasing their voice of protest, justifying the continuation of development projects within sacred ancestral grounds. By paying off Indigenous communities with a minor percentage of the profit margin, renewable energy corporations are continuing the erasure of Indigenous cultures, commonly referred to as green colonialism. Sovereignty means to have control and power over one, or authority of power to self-determination. Indigenous people have the right to sovereignty of their ancestral spaces, to be mindful stewards of these spaces, energy development does not align with that ideology. Energy sovereignty is not the answer to solve this climate crisis. Without the removal of these wind farm turbines, the survival of reindeer herding practices, the ecosystems they support and the communities who depend on their harvest to survive are all in jeopardy within Sampi. Food sovereignty and foodways of the Sami people have become unstable due to green energy development. For the Sami to protect their food sovereignty, energy development initiatives need to be approved with Indigenous veto power in place to come to fair and sustainable solutions to meet green energy goals and lower greenhouse gas emissions.

Other Indigenous Communities impacted by Green Energy Developments

The Kumaonis people in Kumaon Himalayas have been battling India and neighboring Nepal for access to grazing lands which were seized by the state Forest and Revenues Department to gain control over the forested region near their village. By renaming it *The Great Himalayan National Park* the FRD was able to deem the zone protected and block local Indigenous people from using it as grazing pastures as they had been doing for centuries (Singh et al., 2021). The designation implemented new regulations preventing anyone from trespassing on the land. It has sustained and thrived thanks to the reciprocal relationship of care of the Kumaonis for generations provided through selective light use of the forest. The long-term goal of the Indian government was to construct a hydroelectric dam system on the river of the region. The FRD officials insisted the local Indigenous people had overused the resources in the forest and the need to protect it from further encroachment by locals as critical, however, the reality was the complete opposite. The reciprocal relationship locals people shared with the land was one of limited animal grazing, planned replanting's, selective harvesting of seeds for next year's crop, seasonal and sparing morel harvests, rotational harvesting of fodder, and other Indigenous Forest management epistemology and TEK with sustainability in mind (Agrawal, 2005). It was a state sanctioned policy implemented from the start of the 20th century that sought to gain control over the region, so in response the locals created the *Kumaon Forest Grievances Committee.* This led to the state's creation of the Forest Council Rules in 1931. Though the state believed they were protecting the interest of the forest with the new rules, they blocked an entire community of people from their source of food sovereignty; the forest that had sustained them for so long (Agrawal, 2005; Singh et al., 2021). Under state stewardship the forest began to dwindle, overgrowth of brush, introduction of trees not native to the region by the state forest officials pushed out native species impacting wildlife habitat and food sources, causing frequent mudslides. Eventually they built a hydroelectric dam, for the purpose of the land grab. The region became inundated with equipment, scarring the local landscapes. The housing needs the labor forces brought in to build the dam surpassed the village population five times. Improvements in wastewater treatment were required to handle such an impact as existing systems were much too small to accommodate the growth (Agrawal, 2005). This example of exploitation of precious local landscapes that local Indigenous people customarily used for subsistence compares to the Sami struggle. Both are battling for control over their ancestral grounds from industrial development of energy resources.

Another example of Indigenous communities impacted by green energy developments are the peasant studies by A.V. Chayanov, an agricultural economist who worked in the beginning of the twentieth century in Russia throughout the revolution (Chayanov, 1986). In his work he detailed the moral economy or social relations between the elites, and demonstrated how the peasants facing subsistence risks help to create social systems of mutual assistance and tolerable levels of exploitation and inequality. Research from James Scott and E.P. Thompson about the agrarian relationships between moral economy and everyday resistance can be described using subsistence farming struggles. Small producers or farmers who take on the role of protagonists in political ecological stories of land, forest, and fisheries help profit maximization and surplus accumulation for neoliberal capitalist agenda (Robbins, 2020).

The Importance of Reindeer for the Sami and Climate Change

Reindeer herding is a cultural, as well as sustenance practice, of Sami people. It is their main source of food, income, traditional cultural practice, and what allows their survival. The reindeer are harvested at different times of the year at different ages to maximize their value, use, and flavor for consumption. The harvesting of reindeer for the Sami is in line with these seasonal changes to be sure they can sustain this reciprocal relationship. During the spring season, as with most animals, the reindeer calves are born, and selected adult reindeer are taken for harvest to preserve and dry out for consumption through the summer months. In summer, the remaining adult reindeer graze in grassland meadows which gives their meat an undesirable flavor if slaughtered during this period, usually these adults are not taken to harvest until later in the year. It is a selection of calves who are usually chosen to harvest at the end of summer, they have strong tough hides which will make warm clothes for upcoming winter seasons to be worn by herding families during mountain migrations. Additionally, the meat of these young calves is considered a delicacy to the Sami because of its tenderness, much like the differences between beef and veal. By the beginning of fall season, adult reindeer are again selected for slaughter from the herd. Their meats are then smoked to preserve for the migration journey's ahead which sustain the Sami who migrate with their herds. Dressed in warm reindeer leather gear, the nomadic Sami follows behind, but do not steer the reindeers pathways. These animals instinctually follow the most pregnant female out to forage for food in the meadows, across the vast. Reindeer leave the mountain meadows toward the coastline in the summer months to flee the mosquitos and then head back to the mountain meadows by early fall to beat the harsh temperatures of the coast in the winter. The Sami people use the entirety of each reindeer, not wasting it. What is not consumed or used as material can supply trade later. These items include handicrafts carved from their bones, clothing made from their fur and hides, medicine for their sick, iron supplements for their pregnant women, bone marrow for soups, and more (Levy, 2006; DuBois & Lang, 2013). Their meat is harvested again in the winter for consumption throughout the season. Grazing animals play a special role in the health of the ecosystems which sustain them. Their activity promotes new growth and higher levels of diversity in the grasslands, (Agrawal, 2005). Vegetation and biodiversity of grazing area ecosystems cohabitate with the reindeer, dependent upon their seasonal presence as they annually forage during migrations. As the reindeer feed, they help to maintain their own food resources survival, naturally through grazing (Kamerling et al, 2017). The need to protect these places from development is critically important toward maintaining Sami food sovereignty.

Sami People's survival into today is because of the substance they receive from the reindeer but the impacts from implementation of industrial climate change mitigations have threatened their reciprocal relationship and that of their cultures survival into the future. Reindeer nibble the tips of lichen and groundcovers just below the frost of the snowpack where their noses can press through smelling their next meals. Climate change alone has impacted these grazing habits in many ways. If the snowpack is too hard, reindeer cannot sniff out vegetation or break through the snowpack to eat. Without enough snow, insect populations become overwhelming and deter reindeer from grazing these regions. When these possibilities overlap as temperatures increase, reindeer are left with dangerous choices: be attacked by disease carrying mosquitoes to find food or migrate through regions with predators who might attack their young calves and jeopardize the safety of the herd (Skarin et al., 2015). Reindeer are adaptable to many outside influences, but climate change continues to challenge their existence. Living in the arctic zone the lives of the Sami have little environmental impact, working hard to maintain integrity of the ecosystems which they rely on is a focal part of their daily routines.

Neoliberalism and Green Energy Development within Sami Cultural Regions

In the eyes of neoliberal energy development corporations, the Sampi area is perfect to implement new green energy initiatives that will supposedly save the world from fossil fuel dependency (Roland et al. 2012). These multinational corporate interests see the vast wilderness, so dearly respected by the Sami people, ripe for their latest green energy development projects, specifically wind turbine farms, (Mustonen et al., 2010). Out of sight from mainstream societies who inevitably will be using this newly generated energy, windmill turbines have been built throughout Sampi, and are now spread across Sami migratory pathways that have been used for centuries by the reindeer. Even if not all Sami people are herders themselves, their lives are still deeply integrated with reindeer. There have been many research studies on their cultural way of life as herders, impacts that development has had on their sustainable future, and on the impacts green energy development has had on reindeer of the nomadic Sami. These developments spearheaded by neoliberal capitalism as a solution to the climate crisis, directly threaten the food sovereignty of the Sami people, disrupting their cultural practice of coastal fishing, nomadic migrations, depleting reindeer herds

(Karagiannopoulos, 2018). Struggles to fend off developers who seek to extract the region of its natural resources seems never ending. From copper, aluminum, natural gas, gold, oil, uranium, one mining company after another has done exploratory mining in the region for one purpose, to exploit these resources for capital accumulation (Vidmar, 2017). In addition to mining companies, industrial developments such as high-speed rail projects, deforestation, and hydroelectric power projects have been developed within the Sampi region despite protest by the Sami, and other international NGOs (Roland & Löffler, 2012). Ignoring existing regulations agreed to by area nation states, developers and governments alike continue to breach laws in place to protect Sami rights. The impact windmill farms, and turbines have had on the reindeer and local landscapes of the Sami region, the ecosystems reliant upon them all, their direct relationships and how deeply degraded their life would become when turbines entered their spaces, were not understood or rather of value to the state actors and corporations who planned, permitted, and implemented energy infrastructure inside the ancestral Sami regions.

Brief Overview of Wind Power Generation in Sweden, Finland, and Norway

Currently, 38 wind farms are active in Sweden with a total of 698 individual turbines (*Swedes Renewable Energy,* 2021). By 2024 1,100 more turbines are slated to be online at the Markbygden Wind Farm which will become the largest in Europe once active and it will make them the largest onshore wind farm in the world (*Wind Projects*, n.d.). President of the Sami Council, Mattias Åhrén said, "The Swedish state has admitted that the project will destroy at least 25% of the Sami community's winter reindeer herding pastures" but the state has argued that renewable energy is more important than Sami rights (Van Raalte, n.d.). This an example of development projects funded by multinational corporations who are unconcerned with impacts projects of this sort have on local Indigenous communities. The Sami people make up about 10% of the overall population of Sweden, leading Council president Åhrén to suggest the country is more concerned about their economy than the sacred Indigenous Sami life of reindeer herding (Vidmar, 2017). The Reindeer Husbandry Law passed in 1971 was implemented to protect the practice of reindeer husbandry as the Swedish government placed greater controls on the practice, forcing Sami people to obtain licenses and Swedish citizenship to maintain their way of life, disregarding their right to self-determination, and denying their Indigenous sovereignty.

Finland is dealing with much of the same issues as the other regions of Sampi. The country has 11.7 % of electricity production and 9.3 % of electricity consumption covered with wind power. This power is generated from the 962 wind turbines running in Finland as of 2021 (Wind Finland, 2022). Turbines that are now taller than ever before with an average height at above 150 meters tall, with equally wide rotor diameter length (Wind Finland, 2022). Finland has a "feed-in tariff" that incentivizes investments into renewable energy in the country for foreign investments, making it an ideal country to jump into the market. Only growth is projected for the wind turbine industry, and this leaves the Sami looking to their political representatives to do more to combat these corporate entities who have little concern for the Sami.

Currently there are 53 windfarms working in Norway, the country generates 98% of their energy needs from renewable sources. 88% of their energy is generated from hydropower and 10% from wind power; these figures are from the Energy Facts Norway organization who composed and released these statistics from 2021 (Wind Finland, 2022). With the motivation of lowering greenhouse gas emissions 40% by 2030, Norway has collected investments from global energy corporations to further their green energy development sector and expand their implementation of wind turbines after a 3-year moratorium (Buli, 2022). The moratorium was in effect a response to the intense opposition from local Sami herders who have endured devastating impacts. A Norwegian supreme court ruling in October of 2021 found the wind turbine developments of Fosen Vind to be in direct violation of international conventions that protect Indigenous rights (Buli, 2021; UN, n.d.). The operating permits should have never been approved by the Norwegian Minister of Petroleum and Energy, Terje Aasland, who no longer holds this position (Buli, 2021). Aasland was quoted in 2018 before the construction of the project stating, "After careful consideration, the ministry has concluded that there is no basis for stopping the (building) of the Storheia wind turbines," in an interview with Reuters news agency (Karagiannopoulos, 2018). Though the court's ruling deems the wind turbines are directly in violation of Indigenous rights of the Sami, no plan was offered or committed to the dismantling of the turbines. These turbines are still spinning today and continue to disrupt the Sami reindeer herder's way of life, as well as negatively affecting their right to food sovereignty. An obvious contradictory position is how the Norwegian government controls a majority of Fosen Vind, who also operates the windmill projects of the Storheia and Roan wind farms, and still unwilling to state intentions of dismantling the projects (Delbeke, 2022). The current Minister of Petroleum and Energy for Norway, now Marte Mjøs Persen, has suggested the ruling gave no details of what should happen next for these wind farms, and instead suggests there should be a new impact assessment to determine the next steps. This response seems more like a stall tactic to allow the government time, and essentially get away with degrading reindeer habitat and migrations. If the same government who made agreements to protect the rights of their Indigenous people, is the power that controls the corporate energy development in the country, what recourse do the Sami have to hold them to account when supreme court rulings are ignored? As one of the world's leading exporters of energy, Norway has made it their goal to clear a profit on sales of energy to the markets of Europe who are desperate to meet their own lower emission goals as soon as possible. Impending need to find new energy supply has only increased with the recent war between Ukraine and Russia, now squeezing resources everywhere tighter (Kårtveit & Riseth, 2022; Delbeke 2022)

Reindeer and Windmills

There have been extensive studies on reindeer, their migration pathways, and the impacts that wind farms have had on their way of life. Poignant is the scientific research of reindeer behavior led by Anna Skarin, who has been studying impacts from industrialization development on the animals for over 10 years as a local professor at the Swedish University of Agricultural Sciences in Stockholm. Her research has focused on the behavior of reindeer and their impact on the biodiversity of the ecosystems in which they live. She works to quantify the impacts that their foraging during migration has on the health of the environments, and to determine if there are any negative impacts as some politicians have suggested in area nation states when discussing their rights (Skarin, 2014). The vibrations of the low frequency sound made from the spinning blades deters the herds from entering the turbine zones. Disruptions to the lands have great effect on the reindeer, they are extremely sensitive and will avoid areas sometimes for years after the natural habitat has been disrupted by development (Skarin et al., 2015). Despite the research, wind farms continue to be developed in the middle of migration pathways throughout the Sampi regions, across multiple borders. Research shows that once windmills are built within these migration zones, reindeer go great distances to avoid, in some cases avoiding them by miles. This has led to the degradation of the area's entire ecosystem, with hardly any signs of natural landscape left surrounding wind park developments. The roads, vehicles and infrastructure that were required to build the turbines on site, along with ongoing monitoring of these structures has further degraded the regions. These wind turbines are loud, tall structures erected on concrete pads, built in the middle of pristine wilderness zones that have now been destroyed. Construction of these wind parks directly contradicts the UN's International Covenant on Civil and Political Rights, (UN, n.d.). But it also rips away at the right to food sovereignty of the Sami, with no regard to the impact this is having on their survival. There is not much else the Sami can do to sustain life in such harsh terrain. The reindeer, and their reciprocal relationship with them are what enable their livelihoods. Herding provides much more than just a source of food or livestock to the Sami; the reindeer are their life. The neoliberal capitalist agenda to build out the less populated regions in wilderness zones with industrialization development projects has encroached on the sovereign rights of the Sami and created a deep mistrust of the Norwegian government (Delbeke, 2022).

Examples of Sami cultural items related to reindeer include basic functional items such as drinking vessels, to more meaningful items used for medicinal purposes such as reindeer fat used to treat health issues, or in ceremony like Sami Shamen's medicine drums made from reindeer hides, (Demant Hatt & Sjoholm, 2013). The Sami culture hold many artifacts and reliquaries sacred to their culture, especially their shamanic medicine drums, called a "goavddis" by Sami in the Northern regions and "gievrie" by Sami in Southern areas. During the 1500's these items were valued as collectables by outsiders (Schilar, Keskitalo, 2018). The Sami people were often subjects to discrimination due to their contrasting clothing and different way of life. Their ability to thrive in the harsh conditions of the Northern Arctic Zone can be directly attributed to the beautifully crafted clothing they wore. Made from reindeer hide, fur, bone, woven linens, and woolen fibers; dyed bright hues of vibrant red and blue from natural ingredients; adorned with various silver and bronze jewelry (Francis, 2017; Demant Hatt & Sjoholm, 2013). The Sami wear bright colors and noise making jewelry items of utilitarian value in order to be more visual out in the tundra, and to best keep track of their herds. This is one reason bells are commonly seen on reindeer toys, so the Sami can find them when they stray from the herds during migration.

The Various Foodways that are Centric to Reindeer Herding.

The herders have scheduled harvest periods for certain seasons, and at certain ages of the reindeer during specific stages of the migratory periods, all of which help them maintain the health of their herds. From using the reindeer hides for shoes and clothing, to the bones for tools, and their organs and other bodily fluids for medicinal purposes, each part of the reindeer, no matter how small, is cherished. The region has little access to medicine and doctors, so the Sami, to maintain their health, use traditional ecological knowledge as an essential part of that care. Which is deeply entwined with their food resources. From bone marrow broths to heal colds, blood cakes for iron deficiencies, or other ailments, it was the reindeer that provided relief (Dubois & Lang 2013)

Reindeer Herding as an Occupation

Today about 10% of the Sami work as reindeer herders (Dubois & Lang 2013; Roland & Loffler, 2012). Sami are divided into several areas throughout Fennoscandia, between the coast and mountains. Coastal Sami people are referred to as "Sea Sami", or *Sjosamene*; Sami people who reside inland, closer to herding regions are known as "Mountain Sami" or *Fjellsamene Innlandssamene* (Demant Hatt & Sjoholm, 2013). Experiences of the Mountain Sami and their struggles to maintain control of their traditional food sovereignty was part of their daily lives during migration periods. Trekking across cold tundra of the arctic zones, without concern of nation state borders (Zellen, 2009). These herds are not simply livestock for Sami herding families, they have intimate connection with these animals that includes a spiritual reciprocity. The Sami believe the reindeer are their close relatives, who share equal interdependence with nature. It was Jelena Porsanger, the director of the Sami museum in Karasjok who describes the Sami relationship to the reindeer as so close that it is reflective in their language. Where the term "eallit" which means "to live" in English, is rooted in the noun "eallu" which means in English, "herd" such similarity is not accidental (Bogdanova, 2020; Higgins, 2022). The Mountain Sami people live to herd, and herd to live reciprocally.

Conclusion

There needs to be a global agreement that protects Indigenous people from development in their spaces, to preserve their cultural ways of life indefinitely. Without protection many regions like Sampi will no longer exist, they will become fields of development projects hiding in the rural zones of the globe where metropolis populations never have to see, hear, or be impacted by their massive structures. The "not in my backyard" mentality that is prevalent in affluent mainstream society regarding energy infrastructure, needs to recognize that Indigenous communities who survived erasure of colonialism, hold deep value to our future and are not places to extract resources from. Whether that be mining or wind farming. The Sami culture would be barren without the reindeer, survival of the reindeer in the region is unlikely without the Sami, the reciprocity of their relationship is integral to human history. Still to this day the Sami battle to preserve their culture and protect their ancestral spaces from green energy developments. Nation states must stop allowing Indigenous lands to be used for extraction and exportation of resources. Green energy development is not green if it is built on the backs of Indigenous communities.

References

Agrawal, A. (2005) . Introduction: The Politics of Nature and the Making of Environmental Subjects. In *Environmentality: Technologies of Government and Making of Subjects*, New York, USA: Duke University Press. https://doi.org/10.1515/9780822386421-003

Berg, E. (2014). *Sami traditions: Márkomeannu's contribution to the revitalization of Sami food traditions* (Master of Philosophy in Indigenous Studies, The Arctic University of Norway). 22. https://arctichealth.org/media/pubs/295943/10.1.1.897.2063.pdf

Bogdanova, E., Andronov, S., Hossain, K., Dele, R., Praskovia, F., & Lobanov, A. (2020) Food Sovereignty of the Indigenous Peoples in the Arctic Zone of Western Siberia: Response to COVID-19 Pandemic. *International journal of environmental research and public health*, 17(20), 7570 http://doi:10.3390/ijerph17207570

Buli, N. (2021). Norway wind turbines should be torn down, reindeer herders say https://www.reuters.com/business/environment/norway-wind-turbines-should-be-torn-down-reindeer-herders-say-2021-11-12/

Buli, N. (2022). Norway to resume onshore wind power licensing after 3-year break https://www.reuters.com/business/sustainable-business/norway-resume-onshore-wind-power-licensing-after-3-year-break-2022-04-08/

Chaianov, A. V., (1986). *A. V. Chayanov on the theory of peasant economy.* Manchester University Press.

Delbeke, J. (2022). The Impact of the War in Ukraine on Europe's Climate and Energy Policy https://www.euronews.com/green/2022/03/24/what-is-the-impact-of-the-war-in-ukraine-on-europe-s-climate-and-energy-policy

Hatt, E. D. (2013). With the Lapps in the High Mountains: A Woman among the Sami, 1907–1908. University of Wisconsin Pres. muse.jhu.edu/book/22418

DuBois, T., & Lang, J. (2013) Johan Turi's animal, mineral, vegetable cures and healing practices: an in-depth analysis of Sami (Saami) folk healing one hundred years ago.

Journal of Ethnobiology and Ethnomedicine, 9, 57. http://www.ethnobiomed.com/content/9/1/57

Greenson, T., & Weinreb, E. (2019). Why the Supes Denied Terra-Gen's Wind Project; Despite a Series of 11th Hour Concessions from the Company. *North Coast Journal*

Goodyear-Ka opua, N., Hussey, I., & Wright, E. A Nation Rising: Hawaiian Movements for Life, Land, and Sovereignty. (2014). The American Historical Review, 119(5). http://doi.org/10.1093/ahr/119.5.1832

Heaps, D. (2019). Sami land rights and policy-driven recognition threats *University of Washington, June 27*. https://jsis.washington.edu/news/sami-land-rights-and-policy-driven-recognition-threats/

Higgins, C. (2022). 'Our traditions have been criminalized' – the Arctic artists bringing protest to the Venice Biennale. The Guardian. https://www.theguardian.com/artanddesign/2022/mar/31/criminalised-sami-artists-arctic-venice-biennale-loggers-miners-global-heating-culling-protest

Hofverberg, E. (2021) Norway: Supreme court defines extent of Swedish Sami reindeer herder rights. Library of Congress. https://www.loc.gov/item/global-legal-monitor/2021-07-28/norway-supreme-court-defines-extent-of-swedish-sami-reindeer-herder-rights/.

Kaiser, N., Sjölander, P., Liljegren, A., Jacobsson, L., & Renberg, E. (2010). Depression and Anxiety in the Reindeer-Herding Sami Population of Sweden. *International Journal of Circumpolar Health*, 69(4), 383–93. https://doi.org/10.3402/ijch.v69i4.17674.

Kamerling, I., Schofield, J., Edwards, K., & Aronsson, K. (2017). High-resolution palynology reveals the land use history of a Sami renvall in northern Sweden. Vegetation history and archaeobotany, 26, 369-388. http://doi:10.1007/s00334-016-0596-5

Karagiannopoulos, L. (2018, December). Norway to build wind farm despite concerns of reindeer herders. Reuters. https://www.reuters.com/article/us-norway-wind-unidUSKCN1OK1WS

Kårtveit, B., & Riseth, J. (2022, February 2). Norway must stop violating Indigenous rights. Al Jazeera. https://www.aljazeera.com/opinions/2022/2/5/norways-must-stop-violating-indigenous-rights

Levy, J. E. (2006). Prehistory, identity, and archaeological representation in Nordic museums. *American Anthropologist*, 108(1), 135-147. http://dx.doi.org/10.1525/aa.2006.108.1.135

McCune, L. M (2021). Indigenous Food Sovereignty in the United States: Restoring Cultural Knowledge, Protecting Environments, and Regaining Health ed. by Devon A. Mihesuah and Elizabeth Hoover. *Great Plains Research*, 31(2). doi:10.1353/gpr.2021.0031

Mustonen, K., Mustonen, T., & Aikio, A. (2010). Drowning reindeer, drowning homes: Indigenous Sámi and hydroelectricity development in Sompio, Finland.

Morrison, D. (2011). Indigenous food sovereignty: A model for social learning. In *Food Sovereignty in Canada: Towards a just and sustainable food system*. Fernwood Publishing.

Norway's Energy Supply System: *Electricity Production*. (2021). Energy Facts Norway; Norwegian Government. https://energifaktanorge.no/en/norsk-energiforsyning/kraftproduksjon/

Partridge, T. (2016). Water justice and food sovereignty in Cotopaxi, Ecuador. *Environmental Justice*, 9(2), 49–52. https://doi.org/10.1089/env.2016.0003

Peart, N. (2021, September 6). How Indigenous Communities Build Energy Sovereignty. GreenBiz. https://www.greenbiz.com/article/how-indigenous-communities-build-energy-sovereignty

Quentin-Baxter, A. (1999). The UN Draft Declaration on the Rights of Indigenous Peoples-The International and Constitutional Law Contexts. Victoria U. Wellington L. Rev., 29, 85. https://doi.org/10.26686/vuwlr.v29i1.6045

Robbins, P. (2020) *Political Ecology: A Critical Introduction*. John Wiley & Sons.

Pape, R., & Löffler, J. (2012). Climate change, land use conflicts, predation and ecological degradation as challenges for reindeer husbandry in northern Europe: what do we really know after half a century of research?. Ambio, 41, 421-434. http://doi:10.1007/s13280-012-0257-6

Said, E. (1979). Orientalism. Ny. Vintage Books.

Singh, S., Saberwal, V., Singh, N., & Pandey, S. (2021, October 15). Outside In: The Himalayan Land Grab. https://www.nhpr.org/environment/2021-10-15/outside-in-the-himalayan-land-grab

Schilar, H., & Keskitalo, E. C. H. (2018). Ethnic boundaries and boundary-making in handicrafts: Examples from northern Norway, Sweden and Finland. Acta Borealia, 35(1), 29-48. https://doi.org/10.1080/08003831.2018.1456073

Skarin, A., & Åhman, B. (2014). Do human activity and infrastructure disturb domesticated reindeer? The need for the reindeer's perspective. Polar biology, 37, 1041-1054. http://doi:10.1007/s00300-014-1499-5.

Skarin, A., Nellemann, C., Rönnegård, L., Sandström, P., & Lundqvist, H. (2015). Wind farm construction impacts reindeer migration and movement corridors. Landscape Ecology, 30, 1527-1540. http://doi:10.1007/s10980-015-0210-8.

Skarin, A., Sandström, P., Brandão Niebuhr, B., Alam, M., & Adler, S. (2021). Winter and bare ground grazing. Reindeer, Reindeer Husbandry and Wind Power 11, 13. Swedish University of Agricultural Sciences: Nature Conservation Office. https://pub.epsilon.slu.se/26876/1/skarin_a_et_al_220202.pdf

Skarin, A., Sandström, P., & Alam, M. (2018). Out of sight of wind turbines—Reindeer response to wind farms in operation. Ecology and Evolution, 8(19), 9906-9919. https://doi.org/10.1002/ece3.4476

Swedes use a lot of energy – yet emissions are low. The key? Renewable energy. (2021). Sweden. https://sweden.se/climate/sustainability/energy-use-in-sweden

UN declaration on the rights of Indigenous peoples. United Nations https://www.un.org/development/desa/indigenouspeoples/declaration-on-the-rights-of-Indigenous-peoples.html

Van Raalte, J. (n.d.). The Sami reindeer herders of Sweden. Colorado College. https://sites.coloradocollege.edu/indigenoustraditions/sacred-lands/the-sami-reindeer-herders-of-sweden/

Vidmar, L. (2017). Another One Bites the Dust? A Critical Appraisal of the New Draft of the Nordic Saami Convention from the Perspective of Indigenous Rights. Treatises and Documents, Journal for Ethnic Studies issue, 79, 165-183. http://www.dlib.si/details/URN:NBN:SI:doc-GW1OAT9Y

Weinstock, J. (2013). Assimilation of the Sámi: Its unforeseen effects on the majority populations of Scandinavia. Scandinavian Studies, 85(4), 411-430. http://doi:10.1353/scd.2013.0036.

Wind Power in Finland (2022). Finnish Wind Power Association. https://tuulivoimayhdistys.fi/media/tuulivoima_vuositilastot_2021_in_english-2.pdf

Wind Projects. Svevind Powered by Nature; Svevind Group. https://svevind.se/en/projects/

Whyte, K. P. (2016). Indigenous food sovereignty, renewal, and US settler colonialism. In The Routledge handbook of food ethics (pp. 354-365). Routledge.

Zellen, B. S. (2009). On thin Ice: The Inuit, the state, and the challenge of Arctic sovereignty. Rowman & Littlefield. Lexington Books.

Land Acknowledgement

I recognize that the institution where I write this paper, Cal Poly Humboldt, rests on the unceded lands of the Wiyot people. I am writing this acknowledgment to honor their enduring presence throughout these redwoods and coastal mountainous landscape since time immemorial. Please accept this heartfelt invitation to take a deep dive into learning about the richness of their history and vibrant culture.

Acknowledging that this area was developed amidst racial atrocities and forced assimilations, leaving behind generational impacts of trauma and occupation through the exploitation of resources of stolen lands, cultural appropriations, and violent massacres focused on the erasure of Indigenous peoples. I understand that contemporary communities thriving today were built on lands taken from Hupa, Karuk, Mattole, Tolowa, Wailaki, Wiyot, Yurok, and numerous Indigenous Sovereign Nations. This acknowledgment extends to the lands of the Wiyot Tribe, Bear River Rancheria, and Blue Lake Rancheria.

Specifically, my home in Goudi'ni (Arcata), which translates to 'over there in the woods'. The ancestral Wiyot territory stretches from the northern Plhut Gasamuli'm (Little River), to southern Tsakiyuwit (Bear River Ridge), and from Shou'r (Pacific Ocean) to the eastern qus (foothills of the mountains). The river systems of these Wiyot ancestral lands include Baduwa't (Mad River), Hikshari' (Elk River), Wiya't (Eel River), and Girrughurralilh (Van Duzen River.)

As a graduate student and aspiring academic scholar, I acknowledge the indispensable, continuous relationship Indigenous peoples have with their ancestral homelands. The extraordinary landscapes and ecosystems of this Northern California region are a testament to Indigenous stewardship practices working diligently to protect and maintain the pristine beauty of these coastal lands.

In making these acknowledgments, I want to recognize Indigenous peoples' ongoing relationships with these ecosystems of their traditional territories and honor the crucial roles of Tribal nations in preserving these landscapes. As well as their continued struggle to maintain their sovereignty and protect their communities in the future.

The subject of this paper holds special meaning to me as I am a descendant of Sami people who immigrated to Montana from Inari, Finland after fleeing the forced christianization of their villages during the 1800s. Many children were taken by force from their Sami parents and sent to assimilation schools, similar to what was done to Indigenous tribes in the United States during colonization. My great-grandparents were children at the time and met growing up in Belt, Montana, which is the Blackfeet /Niitsitapi (Pikuni) and Gros Ventre/ A'aninin Tribal Nation's territories. Their parents raised cattle in the area and eventually opened a tobacco store. I do not yet know how involved my own family was with settling into the area, but I am researching this lineage now. My grandfather Earl Heikka was a known Western Art sculptor raising his own family in Great Falls, Montana. His artworks depict life at the time in the area from the turn of the century till his death in 1941. This includes hundreds of sculptures from detailed works of local Indigenous people, to mining industry commissions of miners at work to stagecoaches of up to 8 horses. I was born in Hollywood, California, on ancestral lands of the Tongva, Tatviam, Serrano, Kizh, and Chumash Tribal Nations. I include my positionality to share what motivated me to research the issues brought forth in this paper about the plight of the Sami people in the face of the renewable energy revolution and how it is impacting their rights to their sovereignty and self-determination. My goal with this paper is to add to the ongoing scholarly research about the impacts on Sami's cultural ways of life from renewable energy projects implemented within their ancestral lands.

FLOW CREW

ideaFest

ideaFest is a campus-wide event sponsored by the Office of Research & Sponsored Programs, Marketing & Communications, and the Library in collaboration with the Colleges. On May 5, 2023, over 400 students, faculty, and staff from Cal Poly Humboldt gathered to showcase research, performances, digital projects, art, music, and more as part of the 10th annual ideaFest celebration. All were excited for the opportunity to support the development and dissemination of research and creative activities at Cal Poly Humboldt, in person, once again.

Attendees were able to explore the Cal Poly Humboldt Library to share in knowledge and learn about the latest research and creative work that Cal Poly Humboldt has to offer—not to mention partake in charcuterie and refreshments! The following pages are but a small showcase of the many scenes from ideaFest 2023 and many of the posters presented at the event are now availble online in Humboldt Digital Commons. Special thanks to Cal Poly Humbolt and Kellie Jo Brown for photographing the event.

The Marching Lumberjacks kick off the ideaFest 2023 celebration outside of the Cal Poly Humboldt Library main entrance.

The student-organized Marching Lumberjacks take pride in being an exceptionally non-traditional marching band.

Just inside the main library entrance, the music department showcases Humboldt students as soloists, composers, and performers.

Scenes from ideaFest 2023

A Cal Poly Humboldt student helps an attendee enter Maker Space Virtual Reality.

The Cal Poly Humboldt 3D Digital Herbarium aims to spark curiosity with an extensive database and engaging user experience.

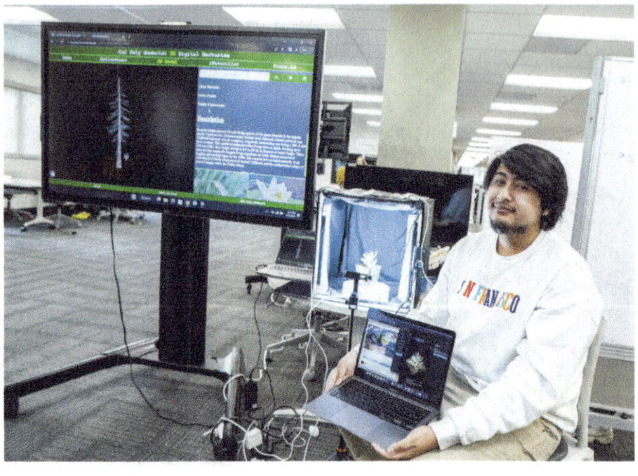

The new Cal Poly Humboldt 3D Digital Herbarium can also be experienced in augmented reality.

Students share a look into Toyon 69: The Sex Issue / La Edición del Sexo, the latest issue of the long-running and student-lead multilingual literary magazine.

Sydney Wells points to her research presentation, Effects of moonlight illumination on animal-vehicle collisions.

Attendees and presenters gather on the second floor under the international flags to share in knowledge and ideas.

Scenes from ideaFest 2023

The Redwood Coast Beer Trail poster, curated by Matthew Gerber, is a self-guided, self-paced wayfinding aide that empowers local and non-local visitors to experience the unique oddities and attributes of the region's craft beer and cider producers.

Undergraduate student Thomas Paulson poses with Cultural Fire Behavior and Effects on Hazel Shrubs, a collaborative effort with Cal Poly Humboldt faculty member Jeffrey Kane and Cultural Fire Management Council staff member Claire Brown.

Tananya Alberts shows off more than just the poster for Identifying Pleistocene Ursus fossils from dental morphology analyses of American black bear (Ursus americanus) and brown bear (Ursus arctos) specimens.

Library student assistants pose at a table displaying a variety of publications, including the ideaFest Journal, produced by The Press @ Cal Poly Humboldt.

Elizabeth Kowalski, Brizeida Mejia Espinoza, and Amar Tojaga are dressed in business best and smiling for the camera with their poster, Assessing and mitigating metabolic response of HEK293 cells to cytotoxic metals using ascorbic acid.

Calla Peltier-Olson poses with her poster entitled "The sense of solidarity": Relational ethics, peer support, and specialized resources for rural and Indigenous social work education program alumni.

Scenes from ideaFest 2023

Authors and editors gather to celebrate the launch of the Humboldt Journal of Social Relations 50th Anniversary Edition: Becoming a Polytechnic.

The Humboldt Journal of Social Relations celebrates its fifty years of scholarship with a display of free back issues.

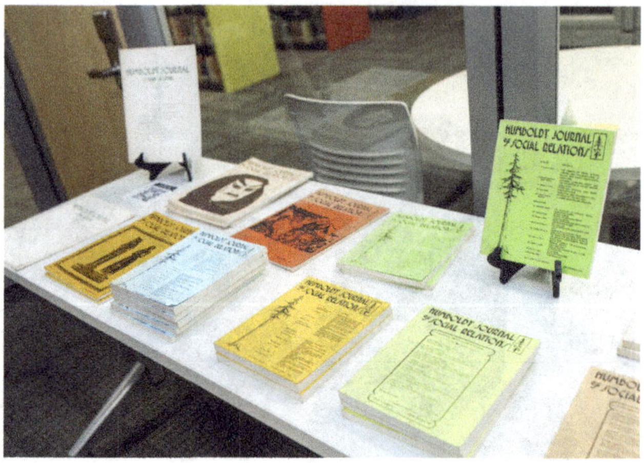

The legacy of Dr. Samuel P. Oliner, founder of HJSR and cofounder of the Aultrustic Behavior Institute, is also celebrated at the journal's launch.

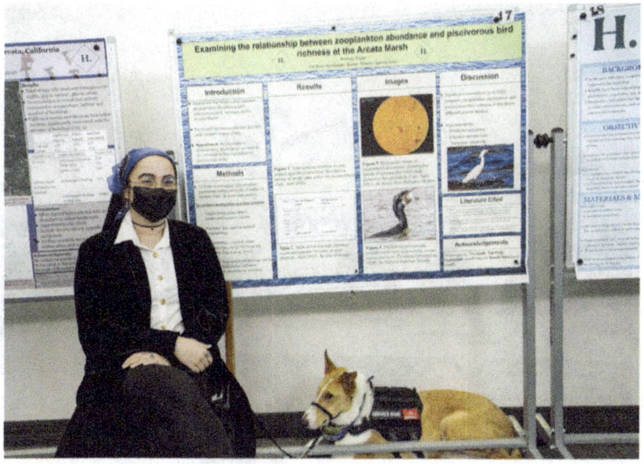

Aminah Aryan and her service dog present Examining the relationship between zooplankton abundance and piscivorous bird richness at the Arcata Mars.

Maia Miglio demonstratess the Humboldt County Homeless College Student - Photovoice Project, a digital humanities gallery project that utilizes various media to teach about and explore the challenges faced by unhoused students.

Attendees collaborate on coloring in an environmental justice roadmap.

Members of the student-run digital humanities creative research team, QTBIPOC Film Coalition, gather with a new zine exploring the art of filmmaking, new media, and immersive/VR technologies from a social justice perspective.

Brooke Culler talks spiders with her poster entitled Plant-Dwelling Spiders Care Where They Live: Higher Diversity and Greater Abundance On Native Plant Species, Specific Plant Types, and in Areas With Vegetation Richness

Fast Fashion, a creative project from Laquita Agwiak, Marco Blancas, and Justin Packham, focuses on the ethics of the fashion industry to expose its devastating cost to the environment with outfits made from trash and found objects.

Made in the USA
Monee, IL
02 November 2026

67377407R00044